Green Innovations for Industrial Development and Business Sustainability

Focusing on the business implications of green innovation, this book describes the sheer impact, spread, and opportunities arising every day, and how business leaders can implement green innovative practices today to realize tangible as well as intangible business advantages in the future.

Green Innovations for Industrial Development and Business Sustainability: Models and Implementation Strategies works as a guide for green innovation that focuses on enterprise applications for those tasked with leveraging green practice adoption to enhance the productivity of their organization. The book offers the ability to understand the latest developments in green innovations and their business applications along with their implications in various domains from manufacturing to marketing.

Front-line decision-makers can use this book as a practical guide for capitalizing on the latest green practices, adoptions, and transformations. Researchers, business leaders, postgraduate, and graduate students will find this book useful with its presentation of state-of-the-art research together with the current and future challenges of building green practice models and applications for organization and business operations.

Information Technology, Management, and Operations Research Practices Series

Series Editors: Vijender Kumar Solanki, Sandhya Makkar, and Shivani Agarwal

This new book series will encompass theoretical and applied books and will be aimed at researchers, doctoral students, and industry practitioners to help in solving real-world problems. The books will help in various paradigms of management and operations. The books will discuss the concepts and emerging trends in society and businesses. The focus is to collate the recent advances in the field and take the readers on a journey that begins with understanding the buzz words such as employee engagement, employer branding, mathematics, operations, and technology and how they can be applied in various aspects. It walks readers through engaging with policy formulation, business management, and sustainable development through technological advances. It will provide a comprehensive discussion on the challenges, limitations, and solutions to everyday problems such as how to use operations, management, and technology to understand the value-based education system, health and global warming, and real-time business challenges. The book series will bring together some of the top experts in the field throughout the world who will contribute their knowledge regarding different formulations and models. The aim is to provide the concepts of related technologies and novel findings to an audience that incorporates specialists, researchers, graduate students, designers, experts, and engineers who are occupied with research in technology-, operations- and management-related issues.

Sustainability, Big Data, and Corporate Social Responsibility: Evidence from the Tourism Industry
Edited By Mohammed El Amine Abdelli, Nadia Mansour, Atilla Akbaba, and Enric Serradell-Lopez

Entrepreneurial Innovations, Models, and Implementation Strategies for Industry 4.0
Edited By Ravindra Sharma, Geeta Rana, and Shivani Agarwal

Multi-Criteria Decision Models in Software Reliability: Methods and Applications
Edited By Ashish Mishra, Nguyen Thi Dieu Linh, Manish Bhardwaj, and Carla M. A. Pinto

Analytics in Finance and Risk Management
Edited By Nga Thi Hong Nguyen, Shivani Agarwal, and Ewa Ziemba

Green Innovations for Industrial Development and Business Sustainability: Models and Implementation Strategies
Edited by Ravindra Sharma, Geeta Rana, and Shivani Agarwal

For more information about this series, please visit: https://www.routledge.com/Information-Technology-Management-and-Operations-Research-Practices/book-series/CRCITMORP

Green Innovations for Industrial Development and Business Sustainability
Models and Implementation Strategies

Edited by
Ravindra Sharma
Geeta Rana
Shivani Agarwal

CRC Press
Taylor & Francis Group
Boca Raton London New York

CRC Press is an imprint of the
Taylor & Francis Group, an **informa** business

Designed cover image: Shutterstock - DreamScope Design

First edition published 2024
by CRC Press
2385 NW Executive Center Drive, Suite 320, Boca Raton FL 33431

and by CRC Press
4 Park Square, Milton Park, Abingdon, Oxon, OX14 4RN

CRC Press is an imprint of Taylor & Francis Group, LLC

Library of Congress Cataloging-in-Publication Data
Names: Sharma, Ravindra (Professor of management), editor. I Rana, Geeta, editor. I Agarwal, Shivani, editor.
Title: Green innovations for industrial development and business sustainability : models and implementation strategies / edited by Ravindra Sharma, Geeta Rana, and Shivani Agarwal.
Description: First edition. I Boca Raton, FL : CRC Press, 2024. I Includes bibliographical references and index.
Identifiers: LCCN 2023057907 (print) I LCCN 2023057908 (ebook) I ISBN 9781032603681 (hardback) I ISBN 9781032604015 (paperback) I ISBN 9781003458944 (ebook)
Subjects: LCSH: Technological innovations--Environmental aspects. I Green technology--Technological innovations.
Classification: LCC HC79.T4 G6875 2024 (print) I LCC HC79.T4 (ebook) I DDC 338.9/27--dc23/eng/ 20240316
LC record available at https://lccn.loc.gov/2023057907
LC ebook record available at https://lccn.loc.gov/2023057908

ISBN: 978-1-032-60368-1 (hbk)
ISBN: 978-1-032-60401-5 (pbk)
ISBN: 978-1-003-45894-4 (ebk)

DOI: 10.1201/9781003458944

Typeset in Times
by MPS Limited, Dehradun

Contents

Preface...vii
Editor's Biography viii
List of Contributors x
Acknowledgements..xii

Chapter 1   A Canonical Correlation Analysis of Green Consumer
Behaviour and Its Contemporary Antecedents....................1

Shalini Reddy Naini and M. Ravinder Reddy

Chapter 2   E-Commerce and Green Packaging: Sustainable
Business Trends..22

Selvi Kannan, Bhakti Parashar, and Amrita Chaurasia

Chapter 3   Synergy of Green Entrepreneurship Among Indigenous
Community: Case for Sri Lanka...40

*Imali Fernando, Hirusha Amarawansha, Sewwandika Gamage,
Jami Perera, and Nipuni Fernando*

Chapter 4   Evaluating Green Human Resource Management Practices in
Family-Controlled Hospitality Business...............................56

Aditi Sharma, Arun Bhatia, and Mridul

Chapter 5   Impact of eWOM on Green Cosmetics Purchasing Intentions:
An Emerging Market Perspective...72

Ramzan Sama and Ravindra Sharma

Chapter 6   Uncovering the Shifting Landscape: A Comprehensive
Analysis of Emerging Trends and Best Practices Adopted in
Green Supply Chain Management...82

V. Harish and Ravindra Sharma

Chapter 7   Green HRM: Eco-Friendly Methods for the Advancement of
Employment...111

Alex Khang, Geeta Rana, and Deepti Dubey

Chapter 8   Green Technology and Its Effect on the Modern World...............120

Deeksha Dwivedi and Shivani Agarwal

Chapter 9   Green Bonds: Accelerating Green Finance Towards
              Sustainable Economic Development 127

              Varsha Gupta and Abhishek Kumar

Chapter 10 Green Innovations Uniting Fractals and Power for Solar
              Panel Optimization 146

              Senthil Kumar Natarajan and Deepak Negi

Chapter 11 Digital Twins in Green Manufacturing: Enhancing
              Sustainability and Efficiency 153

              *A. Mansurali, T. Praveen Kumar, and
              Swamynathan Ramakrishnan*

Chapter 12 Case Studies on Circular Economy Model: Green Innovations
              in Waste Management Industry 165

              S. Senthil Kumar

Index 179

Preface

This book describes how green innovation practices adopted by organizations maintain a balance between nature and itself because of its sheer participation in figuring out and considering its role in achieving the major goals relevant to the changing environmental needs. Raising an alarm is not sufficient to eliminate wrong practices but the inclusion of performing steps needs to be considered and implemented if we want to save our environment for future generations to come. Focusing on the business implications of green innovation, this book describes the sheer impact, spread, and opportunities arising every day, and how business leaders can implement green innovative practices today to realize tangible as well as intangible business advantages in the future.

The strength of the concept of "green innovation" can be affirmed by its spread to many areas like Manufacturing Processes, Human Resource Management, Production Management, Marketing Management, Supply Chain Management, etc. that always envisage thorough involvement in energy saving, pollution prevention, waste recycling, and green product designs. This book will bring forward the green innovative models and their implementation strategies in various business and industry domains for attaining sustainability. The book will also explore the contribution of green innovation to business sustainability and strategic competitive edge and at the same time, it will provide a platform to those enthusiasts who carry a deep interest in green innovation concepts and are eager to find remedial solutions with the help of effective eco-friendly technological advancements. Green Innovation is the key to enabling environmentally sustainable growth as it can lead to a cleaner and safer world. This book also establishes a milestone in understanding global transformational changes occurring in the business world due to green innovation practices and tries to find powerful and environmentally friendly solutions that will improve and streamline business and industry operations. From the mundane to the breathtaking, green practices are already disrupting virtually every process in every industry. As green business practices proliferate, they are becoming imperative for businesses that want to maintain business sustainability.

The book provides front-line decision-makers with a practical guide for capitalizing on the latest green practices adoptions and transformation. The audience also includes researchers and postgraduate and graduate students at the forefront of study on green innovations by presenting state-of-the-art research together with the current and future challenges in building green practices models and applications for organization and business operations.

Happy Reading!

Editors
Dr. Ravindra Sharma
Dr. Geeta Rana
Dr. Shivani Agarwal

Editor's Biography

Dr. Ravindra Sharma is an Assistant Professor in the Himalayan School of Management Studies at Swami Rama Himalayan University, Dehradun, India. He has earned his Ph.D. in Management from Uttarakhand Technical University, India. He holds master's degrees in business administration (MBA) and computer applications (MCA). He also qualified University Grant Commission National Eligibility Test in Management (UGC-NET). Dr. Sharma has more than 16 years of corporate and academic experience. He has conducted many faculty development programs for the national and international audiences. He has published research papers in refereed journals of Emerald, Sage, Springer, IGI Global, and Inderscience. He has published books in the area of employer branding, HR 4.0, Internet of Things (IoT), and artificial intelligence (AI) with reputed publishers like Taylor & Francis Group USA, Nova Science Publishers USA, and IRP Publication House Inda. Dr. Sharma has contributed chapters in different books published in Springer IGI Global, Taylor & Francis, and Palgrave Macmillan. He has presented research papers at national and international conferences. He has also attended a paper development workshop at the Indian Institute of Management in Rohtak, India. He has been honoured as a session chair and keynote speaker at international conferences at various universities. His research interests include employer branding, entrepreneurship, Industry 4.0, and talent management.

Dr. Geeta Rana is an Associate Professor at the Himalayan School of Management Studies in Swami Rama Himalayan University, Jolly Grant, Dehradun, India. She has earned her Ph.D. from the Indian Institute of Technology (IIT, Roorkee) in Management. She has also done a certification course in HR Analytics from the Indian Institute of Management Rohtak (IIMR). She is engaged in teaching, research, and consultancy assignments. She has more than 17 years of experience in teaching and handling various administrative as well as academic positions. She has to her credit over 50 papers published in refereed journals of Emerald, Sage, Springer, Taylor & Francis, Elsevier, and Inderscience publishers. She also presented several research papers at national and international conferences. Dr. Rana has contributed several chapters in different books published by Cambridge UK, Springer IGI Global, and Palgrave Macmillan. She has authored a book titled *Counseling Skills for Managers* published by New Age Publications. She has conducted and attended various workshops, FDPs and MDPs. Dr. Rana is a recipient of many awards. Her research interests include Data Analytics, Human Resource Management 4.0, Knowledge Management, Managerial Effectiveness, Justice, Values, Employer Branding, and Innovation.

Dr. Shivani Agarwal is an Associate Professor, Galgotias University, Greater Noida, India. She has earned her PhD from the Indian Institute of Technology (IIT, Roorkee) in Management. She has completed the executive programme in HR Analytics from IIM Ahmedabad. She is engaged in teaching, research, and consultancy assignments. She has more than 13 years of experience in teaching and in handling various administrative positions. She also presented several research papers at national and international conferences. Dr. Agarwal has contributed chapters to different books published by Taylor and Francis, Springer and IGI Global. She has conducted and attended various workshops, FDPs and MDPs. She is the Book Series Editor of Information Technology, Management & Operations Research Practices, CRC Press, Taylor & Francis Group, USA. She is a guest editor with IGI-Global, USA. Her research interests include Quality of Work life, Trust, Subjective well-being, knowledge management, Employer Branding innovation and human resource management.

Contributors

Shivani Agarwal
Galgotias University
Greater Noida, Uttar Pradesh, India

Hirusha Amarawansha
Department of Management Sciences,
   Faculty of Management
Uva Wellassa University of Sri Lanka
Uva Province, Sri Lanka

Arun Bhatia
HPKV Business School, School of
   Commerce and Management Studies
Central University of Himachal Pradesh
Dharamshala, Himachal Pradesh, India

Amrita Chaurasia
Christ (Deemed to be University),
   Delhi NCR Campus
Ghaziabad, Uttar Pradesh, India

Deepti Dubey
Himalayan School of Management
   Studies
Swami Rama Himalayan University
Jolly Grant, Dehradun, Uttarakhand, India

Deeksha Dwivedi
Department of E-commerce,
   Integration Engineer
Leslie's, Arizona, USA

Imali Fernando
Department of Management Sciences,
   Faculty of Management
Uva Wellassa University of Sri Lanka
Uva Province, Sri Lanka

Nipuni Fernando
Department of Management Sciences,
   Faculty of Management
Uva Wellassa University of Sri Lanka
Uva Province, Sri Lanka

Sewwandika Gamage
Department of Management Sciences,
   Faculty of Management
Uva Wellassa University of Sri Lanka
Uva Province, Sri Lanka

Varsha Gupta
Quantum University
Roorkee, Uttarakhand, India

V. Harish
PSG Institute of Management,
   PSG College of Technology
Coimbatore, Tamil Nadu, India

Selvi Kannan
Victoria University
Melbourne, Australia

Alex Khang
Department of AI and Data Science
Global Research Institute
   of Technology and Engineering
Fort Raleigh, North Carolina, USA

Abhishek Kumar
Quantum University
Roorkee, Uttarakhand, India

T. Praveen Kumar
School of Business and Management
CHRIST (Deemed to be University)
Bengaluru, Karnataka, India

S. Senthil Kumar
Institute of Management Technology
Nagpur, Maharashtra, India

A. Mansurali
Department of Management
Central University of Tamil Nadu
Tamil Nadu, India

Mridul
HPKV Business School, School of
    Commerce and Management Studies
Central University of Himachal Pradesh
Dharamshala, Himachal Pradesh, India

Shalini Reddy Naini
National Institute of Technology (NIT)
Warangal, Telangana, India

Senthil Kumar Natarajan
College of Computing &
    Applied Sciences
University of Technology &
    Applied Sciences
Salalah, Oman

Deepak Negi
Amrapali University
Haldwani, Uttarakhand, India

Bhakti Parashar
Vellore Institute of Technology,
    Bhopal University
Bhopal, Madhya Pradesh, India

Jami Perera
Department of Management Sciences,
    Faculty of Management
Uva Wellassa University of Sri Lanka
Uva Province, Sri Lanka

Swamynathan Ramakrishnan
Amity Business School
Amity University
Dubai, UAE

Geeta Rana
Himalayan School of Management
    Studies
Swami Rama Himalayan University
Jolly Grant, Dehradun, Uttarakhand,
    India

M. Ravinder Reddy
National Institute of Technology (NIT)
Warangal, Telangana, India

Ramzan Sama
Jaipuria Institute of Management
Jaipur, Rajasthan, India

Aditi Sharma
HPKV Business School, School of
    Commerce and Management Studies
Central University of Himachal Pradesh
Dharamshala, Himachal Pradesh, India

Ravindra Sharma
Himalayan School of Management
    Studies
Swami Rama Himalayan University
Jolly Grant, Dehradun, Uttarakhand,
    India

Acknowledgements

The term "Green Innovation" carries a potent and intensified grip that encourages the environment around us toward a pragmatic approach to the need of the hour that emphasizes nurturing our ideas to great heights. The innovations had been a blossom pal to businesses and industries in mitigating tough situations in the past. Adding Green to the term "Innovation" will turn the odds in the favour of industries and business entities that are striving to sustain themselves in the plethora of challenges each passing day.

Today, industries and businesses have realized the price that the whole of mankind is compelled to pay due to negligence and deliberate attempts to ignore the repercussions of an unguided approach to place on the top in the era of the Industrial Revolution. Now, it's appreciated for industries and businesses to adopt green innovative technologies and processes in their business activities. Their flexibility toward involvement in encouraging Green Innovation is a sign of a paradigm shift from dependency on depleting natural resources to renewable energy sources.

This book is an effort to showcase the best green innovative practices adopted by industries and businesses. The included examples, cases, and models have facilitated the technologies to grow exponentially and set an example of how the competitors in the same business framework join hands to make a more conducive environment toward adopting green innovative technology and processes. Considering globalization in the context, businesses around the globe have been interconnected with each other paving a path for green innovation to enter the stream of business processes around the world.

This book resonates with the efforts of every individual associated and the project work was completed successfully with the presence of the Almighty in the form of constant wisdom derived from his blessings.

The book's subject matter has been possible due to the sheer interest of academic colleagues from around the world. We would like to express our gratitude and appreciation for the contribution of the authors.

Great support from academic and industry friends and colleagues made this book successful. The multi-facet support includes review and re-review of chapters, suggestions for revisions, modifications, and critical comments to make the chapters more thorough and meaningful. We acknowledge their tremendous support and are thankful for their valuable comments.

We thankfully acknowledge all the support, inspiration, and motivation we received from our faculty colleagues. We would remain indebted to their outstanding intellectual efforts. Our gratitude is due to Dr. Vijay Dhasmana the Chancellor of Swami Rama Himalayan University, Jolly Grant, Dehradun, Uttarakhand, India; Dr. Rajendra Dobhal, the Vice Chancellor of Swami Rama Himalayan University; and Dr. Vijendra Chauhan, Director General Academic Swami Rama Himalayan University for their faith, trust, and responsibility that empowered us and for extending full support and being a constant source of inspiration and pillars of strength throughout the pursuit.

Our Special thanks goes to Dr. Santosh Rangaker Indian Institute of Technology, Roorkee, Uttarakhand, India, and Dr. H.K. Dangi Delhi School of Economics, University of Delhi, India for their constant encouragement and support in this endeavour. We thank our prospective readers in advance for they would be a source of improvement and further development of this book.

We express our thanks to our publisher CRC (Taylor & Francis Publication, USA) and the entire editorial team who lent their wonderful support throughout and ensured the timely processing of the manuscript and bringing out the book.

Finally, it would not be justified if we did not acknowledge the continuous and unstained support from our families and their belief in us that facilitated accomplishing our task and bringing forth this compiled project in a book form for scholars, readers, and content enthusiasts.

Editors
Dr. Ravindra Sharma
Dr. Geeta Rana
Dr. Shivani Agarwal

1 A Canonical Correlation Analysis of Green Consumer Behaviour and Its Contemporary Antecedents

Shalini Reddy Naini and Dr. M. Ravinder Reddy

1.1 INTRODUCTION

Over the past few years, consumers' consumption worldwide has increased greatly, leading to natural resource depletion and environmental damage, consequently accelerating global warming (Chen and Chai, 2010). Joshi and Rahman (2015) state, that to mitigate this threat, several nations are attempting to scale back their environmentally harmful commercial practices, which has sparked the rise of green development, which promotes eco-innovation and environmentally friendly consumerism. Eco-innovation involves adopting sustainable practices in every phase of creating products, and green consumption is consumers' consideration of the environmental impact during the purchase, use and disposal of the products. The corporate environment has experienced that stakeholders and customers are increasingly open to and mindful of sustainability issues – especially the destruction of forests, pollution, and temperature rise – and they are now shifting to ethical buying practices for the sake of future generations (Jaiswal and Singh, 2018), and focusing on sustainability due to stricter environmental regulations (Panda et al., 2020). India, one of the economies that is growing the fastest worldwide, is experiencing a rise in airborne pollutants and the depletion of its plentiful resources due to this expansion (Joshi and Rahman, 2016). Additionally, businesses in China and India are beginning to realise how conscious of the environment consumers are and have consequently started to restructure their business models to support these practices (Sharma et al., 2022). The Indian government has introduced various green initiatives like Swach Bharat and green tax to promote sustainability (Panda et al., 2020; Sharma et al., 2022). Sustainable development and environmentalism have been on the rise in India, with more consumers becoming environmentally aware and also curious about their consumption habits and how they impact the environment, leading to extensive research in the green marketing field to promote

DOI: 10.1201/9781003458944-1

1

green consumerism (Uddin and Khan, 2018). Green consumption involves a green purchase, which is the purchasing of eco-friendly products, and not considering the products that harm the ecology (Chan, 2001). Sustainable purchasing is vital to society, as unplanned consumption severely damages the environment, and the consumers behind these purchases possess the capability of avoiding this damage by approximately 40% (Grunert and Juhl, 1995). Green purchasing is generally measured using green purchase intention (GPI) and green purchase behaviour (GPB); and a wide range of individual and situational factors affect environmentally friendly purchasing (Joshi and Rahman, 2015). Panda et al. (2020) state, given that consumer tastes are cyclical in nature and that selling eco-friendly items can be challenging, it is crucial to comprehend what customers want and how they make their choices in regard to green products. Based on studies, the majority of Indian customers have exhibited a favourable mindset towards items and gadgets featuring energy-efficient, recyclable packaging, cruelty-free, and fair trade (Jaiswal and Kant, 2018), Hence, understanding the consumer green behaviour is significant, and it is vital to explore the factors that influence the green purchase phenomenon in India, as the consumers and different product and service marketers will benefit from the information that directs the consumer's green-related behaviours (Bailey et al. 2016). In terms of the requirement for environmental sustainability, the study of the factors that affect sustainable behaviour has exploded over the past ten years, with a particular focus on ecological purchasing (Do Paço et al., 2019). However, developing nations like India still have an insufficient comprehension of the factors that influence GPI and GPB.

The factors explored so far were found to either encourage or discourage green purchasing, and the role these determinants play in consumption varies in different contexts and cultures (Joshi and Rahman, 2015). Therefore, it is important to analyse the magnitude of the relationship between the factors and green purchasing measures in the Indian context, as the studies related to this area are little scant and unclear in Asian developing countries (Jaiswal and Kant, 2018) and direction of the relationship between the various antecedents with GPI and GPB separately is found to be different in various settings (Joshi and Rahman, 2015). Researchers in the past used a variety of strategies to pinpoint the many variables influencing consumers' attitudes and behaviours towards eco-friendly items in different settings. Thus, the major influence factors aid marketers in market segmentation and GPI and GPB maximisation. The interplay of various significant factors with GPI and GPB in different contexts is clearly explained in the following section.

1.2   THEORETICAL BACKGROUND

Numerous factors have been proposed in a variety of situations for researching how they affect GPI and GPB, which are discussed below as explanatory or predictor variables. Altruism (ALT), which refers to a concern for the welfare of society (Stern et al., 1993) is positively correlated with green consumer behaviour in USA and Portugal (Straughan and Roberts, 1999; Akehurst et al., 2012); with GPB in the UK (Padel and Foster, 2005), India (Uddin and Khan, 2018), and Portugal (Akehurst et al., 2012); with GPI in India (Akehurst et al., 2012; Panda et al., 2020),

China (Ali et al., 2020; Wang et al., 2020a), and Pakistan (Li et al., 2020). Interpersonal influence (IPI) is the way one convinces or persuades others and it determines the behaviour of an individual (Uddin and Khan, 2018). It is positively associated with GPI in Taiwan (Chang, 2015), and India (Malik et al., 2017); GPI and GPB in Taiwan (Chang and Chang, 2017); with GPB in Pakistan (Zafar et al., 2020), and India (Khare et al., 2013); with green behaviour in India (Khare, 2014). Green behaviour (GB) also known as pro-environmental behaviour or ecologically conscious consumer behaviour explains an individual's behavioural orientations that minimise the harm to the environment like re-using, re-cycling, conserving water, reducing waste and green activism participation (Akehurst et al., 2012; Mishal et al., 2017). It is also the extent to which a consumer purchases products that are believed to have a good impact on the environment (Straughan and Roberts, 1999). It is positively correlated with GPI and GPB in Portugal (Akehurst et al., 2012), and Malaysia (Mas'od and Chin, 2014); with GPB in India (Mishal et al., 2017). Green culture (GC) refers to society's collective environmental behaviour for helping nature and solving climate issues, it is positively associated with EA in Nigeria (Ogiemwonyi et al., 2020a,b); with GB in Pakistan (Afridi et al., 2023) and also in Malaysia and Nigeria (Ogiemwonyi et al., 2020b), and no studies so far have investigated its relationship with GPI and GPB. Green habit (GH) is automatically performing an eco-friendly behaviour and it is positively associated with GPI and GPB in Indonesia and Malaysia (Ghazali et al., 2018); with GPB in Bangladesh (Siddique, 2021). Green awareness (GA) refers to consumers' consciousness that their green product consumption will contribute great value to the environment (Rahmi, 2017). Its relationship with GPI and GPB is studied in Indonesia and proved to be insignificant (Rahmi et al., 2017), and is positively correlated with GPI in Indonesia (Alamsyah et al., 2020; Lestari et al., 2021), Pakistan (Rizwan et al., 2014; Mansoor and Noor, 2019), and Nigeria (Ayodele et al., 2017), it also sometimes showed an insignificant and weak effect in Pakistan (Rizwan et al., 2013). It shows a positive correlation with green purchase decisions in Bangladesh (Nekmahmud and Fekete-Farkas, 2020) and is positively related to GPB in Malaysia (Sh. Ahmad et al., 2022). Subjective norms (SN) are how individuals perceive pressure from their social peer groups for indulging or not indulging in green behaviour (Taufique and Islam, 2021) it can also be defined as a group evaluation of the appropriateness of an individual's behaviour exerting a social pressure in him/her (Sun and Xing, 2022). It is positively correlated with GPI and GPB in China (Xu et al., 2022), India (Sethi et al., 2018) and Iran (Nejati et al., 2011), and negatively correlated with GPI and GPB in Portugal (Sousa et al., 2022); positively associated with GPI in China (Liu et al., 2020; Sun and Xing, 2022), Turkey (Albayrak et al., 2013), Vietnam (Vu et al., 2022; Duong et al., 2022; Nguyen et al., 2017), Jordan (Al Zubaidi, 2020), Algeria (Alalei and Jan, 2023) and India (Sreen et al., 2018; Bhatt and Bhatt, 2015). It is negatively correlated with GPI in Taiwan (Ruangkanjanases et al., 2020); positively related to green purchase decisions in Malaysia (Noor et al., 2017); negatively associated with GPB in Vietnam (Nguyen et al., 2018). SMM is the use of online applications, and media in marketing activities, which aim to promote communication, collaboration and content sharing (Erkan and Evans, 2016; Sun and Wang, 2019). It is positively correlated with GPI in China (Zhao et al., 2019; Sun and Xing, 2022; Sun et al., 2022), Romania and Hungary (Pop et al., 2020; Nekmahmud et al., 2022), and Jordan (Al-Gasawneh and Al-Adamat, 2020). The social media usage and E-WOM relationship with GPI and GPB were studied in Indian millennials and showed significant results (Jain et al., 2020). The effect of social media marketing communications on GPI and GPB was studied in Pakistan (Adam and Ali, 2022). The perceived social media marketing is positively correlated with behavioural intentions and GPB in Pakistan (Ch et al., 2021). It is positively associated with attitudes towards green products and willingness to pay in India (Gupta and Syed, 2022). The social media context relationship with a sustainable purchase decision was studied in China (Zafar et al., 2021), with green consumption intentions studied in Vietnam (Chi, 2021), with GPI in China (Luo et al., 2020) and the USA (Bedard and Tolmie, 2018), along with GPI in Pakistan (Zahid et al., 2018). Environmental attitude (EA) is defined as a person's ability to assess the condition of the environment with some agreement or disagreement (Uddin and Khan, 2018). It is positively correlated with GPI in Indonesia (Kusuma and Handayani, 2018; Indriani et al., 2019), India (Sreen et al., 2018; Joshi et al., 2021), USA (Ahmed et al., 2023), Jordan (Al-Quran et al., 2020), Cambodia (Liao et al., 2020), Taiwan (Huang et al., 2014; Rahimah et al., 2018; Thi Tuyet Mai, 2019), Iran (Irandust and Bamdad, 2014; Naalchi Kashi, 2020), Malaysia (Mei et al., 2012; Hasnah Hassan, 2014), Pakistan (Siyal et al., 2021; Wang et al., 2022), Bangkok (Munamba and Nuangjamnong, 2021), Yogyakarta (Baiquni and Ishak, 2019), Vietnam (Nguyen et al., 2017; Thi Tuyet Mai, 2019), Turkey (Albayrak, 2013) and China (Liu et al., 2020; Wang et al., 2020a,b); with GPB in Egypt (Mostafa, 2007), Malaysia (Tan and Lau, 2011; Noor et al., 2012; Sh. Ahmad et al., 2022), India (Joshi and Rahman, 2016; Chaudhary and Bisai, 2018; Uddin and Khan, 2018), Korea (Han et al., 2020), China (Cheung and To, 2019), South Korea (Ok Park and Sohn, 2018), and Vietnam (Nguyen et al., 2018); with GPI and GPB in UK (Kanchanapibul et al., 2014), India (Mishal et al., 2017; Yadav and Pathak, 2017), South Africa (Dilotsotlhe, 2021), Indonesia and Malaysia (Ghazali et al., 2018), Pakistan (Ali et al., 2011; Qureshi et al., 2022), China (Chan, 2001; Xu et al., 2022), Portugal (Fontes et al., 2021) and Bangladesh (Zahan et al., 2020; Siddique et al., 2021). Hence, it is clear that EA's relationship with green consumer behaviour has been widely studied and most of the studies concluded that it is significant in studying the GPI and GPB, acts as an important mediator between the independent and the GPI/GPB variables, and should be included along with other variables. Perceived environmental knowledge (PEK) is the consumer's knowledge about the impact of usage of a product on the ecology and whether the product is produced in an eco-friendly manner; it is positively correlated with GPB in India (Joshi and Rahman, 2015; Uddin and Khan, 2018), Hong Kong (Lee, 2010), Ghana (Martins, 2022), Thailand (Chaihanchanchai and Anantachart, 2023), Egypt (Mostafa, 2007), Malaysia (Noor et al., 2012; Tan, 2011; Ali, 2021), Thailand (Chaihanchanchai and Anantachart, 2023); with GPI in Indonesia (Indriani et al., 2019; Kusuma and Handayani, 2018; Hariyanto and Alamsyah, 2019; Fabiola and Mayangsari, 2020; Nia et al., 2018; Veer et al., 2022), China (Chen, 2013; Lee, 2017), Sudan (Mahmoud et al., 2017), Korea (Lee, 2017), Malaysia (Goh and Balaji, 2016;

Wang et al., 2020b; Hamzah and Tanwir, 2021), Taiwan (Moslehpour et al., 2022), and India (Tudu and Mishra, 2021); with GPI and GPB in UAE (Khaleeli et al., 2021), Indonesia and Malaysia (Ghazali et al., 2018), Bangladesh (Zahan et al., 2020), UK (Kanchanapibul et al., 2014), India (Sharma and Foropon, 2019), Hungary (Naz et al., 2020), Taiwan (Kamalanon et al., 2022) and Indonesia (Arisbowo and Ghazali, 2017). It is not correlated with GPI in Indonesia (Eles and Sihombing, 2017; Qomariah and Prabawani, 2020), Pakistan (Ansari and Siddiqui, 2019), with GPI and GPB in Srilanka (Samarasinghe and Samarasinghe, 2013). Therefore, the role of PEK in analysing the GPI/GPB showed inconsistent results across various settings and must be examined clearly along with other variables.

Hence, It is clear that most of the studies examined the GPI and GPB separately, and very few countries like Portugal, Malaysia, China and Indonesia have studied GPI and GPB altogether in their research. As per the researcher's knowledge, none of the studies has investigated the relationship between this linear combination of a wide range of predictors and the linear combination of GPI and GPB, i.e., the overall green purchasing in India. This study incorporated the contemporary (SMM, GC, GH, AlT, GA, GB) and noteworthy (EA, PEK, IPI, SN) variables in the green marketing literature and analysed their relationship with GPI and GPB. By examining the influence of these ten variables on Indian consumers' green purchasing decisions, our study aims to close this research gap.

1.3   METHODOLOGY

The research is an exploratory study. It is also a quantitative one using the random sampling technique. A total of 510 responses were collected online from September to November 2022 from the Indian population through various sources like LinkedIn, Twitter, Facebook, etc., out of which 506 responses were valid.

The survey questionnaire consists of demographics like gender and age group, and socio-economic items like occupation and education. The main part of the questionnaire consists of the items taken from several studies. The items for GA were taken from Chen et al. (2018); PEK (Mostafa, 2006); GB (Roberts, 1996; Ahn, Koo and Chang, 2012's GPB construct and Lee, 2014's recycling participation items); EA (Mishal et al., 2017); IPI (Lee, 2009); GPI (Chan, 2001); ALT (Stern et al., 1993); SN (Sun and Wang, 2019); GPB (Lee, 2009); GH (Verplanken and Orbell, 2003); SMM (Sun and Wang, 2019); GC (Ogiemwonyi et al., 2020a).

The analysis is conducted at three levels. First, demographic analysis is conducted to identify the distribution of the study sample. Second, Pearson correlation analysis is performed to analyse the correlations between the variables. The final stage of the analysis adopts a multivariate approach. The variables are drawn together in the application of canonical correlation analysis (CCA) to investigate the relationship between the variable sets. Specifically, it inspects the association between two endogenous variables, GPI and GPB, with ten exogenous variables considered as green factors or determinants. CCA is an extension of multiple regression (Mishra, 2015). As in multiple linear regression, it is not confined to a single criterion variable but is concerned with the relationship between the sets of predictor variables and dependent variables (Green et al., 1966). It also determines the magnitude and nature of the relationship between the two sets of variables by measuring the relative contribution of each variable to the canonical function obtained (Alpert and Peterson, 1972).

1.4 RESULTS

1.4.1 Demographic Analysis

The descriptive results reveal that the majority of the sample is in Gen Y (67%), and are postgraduate (47.8%) males (63.4%) who are mostly employed (48%). In terms of age distribution, 20.6% were Gen Z, 67% represented Gen Y and 12.5% constituted Gen X. The gender-wise distribution states that 63.4% of respondents were male and 36.6% were female. According to occupation, 48% were employees, followed by students (37.9%), others (9.9%), and own business (4.2%); In terms of the highest education level, postgraduates were 47.8%, trailed by above post-graduation (31.6%), graduation (17%) and Intermediate/high school (3.6%).

1.4.2 Correlation Analysis

The main purpose of the research study is to observe the underlying principle of the assumption that GPI and GPB are dependent upon the ten factors. So as an initial step, the Pearson correlation analysis is performed to examine the correlation between the GPI and GPB and the correlation between each of these two variables with the ten independent variables (Table 1.1). This analysis is carried out to test the null hypothesis that the population correlation coefficient is zero and to lay the groundwork for investigating if there is a correlation between the predictor and criterion variables (Green et al., 1966). The findings show that, with the exception of the ALT with the IPI and SN, all correlations are different from zero at a 5% level of significance.

There exists a high correlation between the GPI and GPB. There are also significant correlations between the GPI and all the ten factors, in the decreasing order of EA, IPI, PEK, GB, GH, SN, SMM, GC, GA, and ALT. The correlations between GPB and the factors are also significant, and it is strongly associated with IPI, PEK, GB, EA, and GH. The correlations of GPI/GPB with ALT, GA and GC are generally weaker, though significant. From the Pearson correlation results, it seems logical to conduct CCA.

1.4.3 Canonical Correlation Analysis

A multivariate technique termed canonical correlation analysis (CCA) measures the strength of the overall relationship between a linear combination of exogenous and endogenous variables (Mai and Ness, 1999; Mishra, 2015). The number of predictors is ten and the dependent variables are two, so the analysis considers the smaller one, and concludes the utmost number of functions, hence deriving two functions in this case.

TABLE 1.1
Correlations

	ALT	GB	GC	GA	PEK	IPI	SMM	SN	EA	GH	GPI	GPB
ALT	1											
GB	.180**	1										
GC	.424**	.285**	1									
GA	.353**	.267**	.636**	1								
PEK	.190**	.535**	.373**	.342**	1							
IPI	.035	.504**	.149**	.118**	.565**	1						
SMM	.096*	.307**	.176**	.189**	.339**	.491**	1					
SN	.084	.354**	.269**	.210**	.380**	.487**	.569**	1				
EA	.263**	.324**	.477**	.505**	.364**	.347**	.374**	.432**	1			
GH	.209**	.399**	.178**	.150**	.386**	.415**	.253**	.245**	.308**	1		
GPI	.192**	.451**	.308**	.294**	.456**	.471**	.368**	.421**	.526**	.445**	1	
GPB	.159**	.531**	.315**	.275**	.547**	.573**	.419**	.465**	.512**	.467**	.638**	1

Notes
** Correlation is significant at the 0.01 level (2-tailed).
* Correlation is significant at the 0.05 level (2-tailed).

TABLE 1.2
Wilks Statistic

	Correlation	Eigenvalue	Wilks Statistic	F	Num D.F	Denom D.F.	Sig.
1	.775	1.502	.391	29.564	20.000	988.000	.000
2	.144	.021	.979	1.171	9.000	495.000	.311

In common, the function's interpretation can be done by observing the values of canonical weights/loadings/cross-loadings. Hair et al. (1998) suggest that cross-loadings are better to loadings, which are then finer to weights. However, in reality, the researcher will be inhibited by the features and form of output generated by the software and options exercised by the analyst (Mishra, 2015). As the canonical loadings in SPSS give the structure correlations, that is it measures the correlation of each variable in the function with the linear combination of variables in the set, the interpretation of CCA in this study is based on the loadings.

Table 1.2 presents the results of the Wilks statistic that is employed for assessment of the analysis. The multivariate tests of significance determine whether the predictor and criterion variables significantly correlate. The null hypothesis assumes no statistically significant relationship which means there is no correlation between the variable sets (correlation is zero), whereas the alternative hypothesis is that the association is statistically significant that is both sets of variables have a correlation (correlation is not zero/different from zero). The findings show that, for function 1, H0 is rejected at a 5% level of significance, demonstrating a statistically significant connection between the exogenous and endogenous variable sets. The canonical correlation of 0.30 or less is treated as trivial (Mishra, 2015), here the canonical correlation for function 1 is 0.775, which indicates a high correlation, and 0.144 for the second function. The question of relevance is whether the correlations are significant. Since the second correlation function's p-value is more than 0.05, it is not. Moreover, as the value of Wilks' lambda is less in the case of function 1 as compared to function 2, the explanatory power in function 1 is more compared to function 2.

The canonical loadings are given in Table 1.3. These loadings are alike the factor loadings and tell the importance of variables. The loadings of function 1 are higher than those of function 2. For all these reasons along with the statistical evaluation, the second function is not statistically significant and has a minor contribution and practical value. So the subsequent interpretation focuses on the first function.

The criteria variables and the independent variables exhibit strong relationships for function 1. The variation explained by the independent variables, calculated by squaring the correlations, is 73.4% for GPI and 88.9% for GPB.

The results indicate that, in descending order of significance, the IPI, EA, PEK, GB, GH, SN, SMM, GC, GA, and ALT have the strongest correlations. Variables with correlations of 0.30 (9% of variance) and above are usually interpreted as part of the variate, and variables with loadings below 0.30 are not (Meloun and Militký, 2011), here the ALT variable correlation value is below 0.30, so it does not necessarily contribute to the variate. The overall high correlation of 0.775 shows a relationship between high ratings on the predictors and greater GPI and GPB levels.

TABLE 1.3
Canonical Loadings

	Function	
Variables	1	2
Independent		
GC	−.444	−.183
GA	−.401	−.373
PEK	−.725	.344
IPI	−.757	.423
EA	−.734	−.513
GH	−.650	−.177
ALT	−.244	−.404
GB	−.709	.266
SMM	−.566	.111
SN	−.635	.008
Dependent		
GPI	−.857	−.515
GPB	−.943	.332

Table 1.4 explains the redundancy analysis of the CCA. Mishra (2015) states "redundancy index is the amount of variance in a canonical variate explained by the other canonical variate in the canonical function". These are the averages of the squared loadings for a particular variate and are the measures of possible predictive ability. A redundancy index, suggested by Stewart and Love in 1968, gauges how well the explanatory variables can account for the variance in the outcome variables. Here the amount of variance explained by one set of variables by another set of variables is maximum in function 1 compared to the second function. The total variance of the ten factors explained by the independent variate is 37.1% and the total variance of GPI and GPB explained by the dependent variate is 81.2%. The variance explained between independent and dependent canonical variables for function 1 is 98.6%, and function 2 is 1.4%, which are calculated by the eigenvalues in Table 1.5. Therefore, around 99% correlation is explained by this canonical correlation in function 1, which further supports the exclusion of the second function. The results reported and interpreted for function 1 are shown in Figure 1.1.

TABLE 1.4
Proportion of Variance Explained

Canonical Variable	Set 1 by Self	Set 1 by Set 2	Set 2 by Self	Set 2 by Set 1
1	.371	.222	.812	.488
2	.101	.002	.188	.004

TABLE 1.5
Eigenvalue and Variance

	Eigenvalue	Variance %
1	1.502	98.6%
2	.021	1.4%
Total	1.523	100%

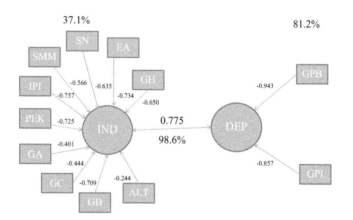

FIGURE 1.1 Function 1 (Canonical correlation relationship between the independent variable set and the dependent variable set). Note: IND, Independent; DEP, Dependent; GPB, Green purchase behaviour; GPI, Green purchase intention; ALT, Altruism; GB, Green behaviour; GC, Green culture; GA, Green awareness; PEK, Perceived environmental knowledge; IPI, Interpersonal influence; SMM, Social media marketing; SN, Subjective norms; EA, Environmental attitude; GH, Green habit.

Source: Author's own.

1.5 SUMMARY AND CONCLUSION

The study aims to understand Indian green purchasing in terms of GPI and GPB, and its determinants as a measure of ten factors such as GB, SMM, SN, ALT, GH, IPI, GA, GC, PEK and EA.

The descriptive results reveal that the majority of the sample is in Gen Y (67%), and are postgraduate (47.8%) males (63.4%) who are mostly employed (48%).

The Pearson correlation analysis revealed that there exists a significant correlation between the GPI and GPB. Each of these variables also has a significant positive association with each of the ten determinants.

The advantage of the CCA is that it considers the simultaneous interaction between all scales, as opposed to the univariate analysis of scales. It is an extension of multiple regression analysis. The results indicate a statistically significant association between the independent and dependent variable sets, consistent with the literature. The results indicate that green purchasing is strongly associated with the behavioural and cognitive aspects along with psychographic and external marketing cues, which is quite interesting. The results are consistent with the contemporary prominence on augmenting green consumer behaviour, which is anticipated as an objective and a prime element in the attainment of a sustainable world for developing countries, through those determinants concerned with consumers' psychological (EA, ALT, SN, GC, GA) along with the behavioural (GH, GB), external marketing cue (SMM) and stimuli (IPI), cognitive (PEK) variables.

The results show that the IPI, EA, PEK, GB and GH are comparatively more powerfully associated with GPI and GPB, than the SN, SMM, GC, GA, and ALT. The ALT variable does not necessarily correlate with the IPI and SN and has a loading of less than 0.30. The high loading on GPB in comparison with GPI, implies that the predictors are highly contributing to the consumers' eco-friendly purchase behaviour and their intentions are turning into actions. Consequently, the message to Indian marketers is that they need to focus on increasing peer group interactions on different platforms like LinkedIn, Facebook, and Instagram to encourage positive word-of-mouth publicity. The contests and activities that encourage group discussions must have been opted for. Indian consumers have a positive and favourable attitude towards the purchase of green products, so the knowledge related to green product packages, labels, phrases and symbols must be spread around to help consumers in selecting the greens. The knowledge about recycling and other environmental issues will help society in developing a sustainable attitude and encourage green consumption. The government and NGOs should make the public aware of the companies that are indulging in ecologically irresponsible activities and help them make a choice. Everyday green programmes and activities involving planting trees, recycling, categorising garbage, and purchasing bio-degradable products must be initiated. The green habits that involve purchasing eco-friendly products frequently and on a daily, weekly, and monthly basis must be encouraged by marketers and retail stores by offering different offers, discounts and coupons.

The use of social media for searching and sharing opinions related to sustainable products has increased in recent days, so marketers should concentrate on social media advertising to promote their products and target the audience. Search engine optimisation must also be utilised for spreading eco-friendly products and green messages. The role of subjective norms on GPI and GPB is also high. The government should introduce group community projects to enhance the green culture and awareness among the public, which helps in reducing environmental degradation and together behave in an eco-friendly way. The sales promotions designed by the marketers must introduce group discounts and community coupons to encourage broader green product consumption. The role of altruism is very low in Indian green consumption behaviour. The need and benefits of having concern for the well-being of future generations and society must be spread across the globe through different movies, web series, books, campaigns, comics, packages and advertisements.

1.6   FUTURE RESEARCH DIRECTIONS

The study concentrates on general green consumer behaviour, for no specific product category; future researchers can elaborate on this work and focus on specific areas like apparel, food, and electronics to conduct comparative studies and check the Indian GPI and GPB in multiple categories. The CCA can be followed by other multivariate techniques like exploratory factor analysis, and conjoint analysis to analyse the behaviour in detail. This study is an exploratory and longitudinal one, which can be elaborated by conducting cross-sectional data collection to address consumer behaviour dynamics. The determinants in this study are chosen according to their significance to the green marketing area and few of them are contemporary and have not been explored in the Indian context. The potential researchers may explore other variables along with the ones in this study to examine their interplay of cause-and-effect relationships.

REFERENCES

Adam, M. A. and Ali, S. M. (2022), "Influence of social media marketing communications on young consumers' attitudes and purchase intention", *GMJACS*, Vol. 12 No. 1, pp.16–16.

Afridi, S. A., Zahid, R. M., Khan, W. and Anwar, W. (2023), "Embracing green banking as a means of expressing green behavior in a developing economy: exploring the mediating role of green culture", *Environmental Science and Pollution Research*, pp.1–11. doi: https://doi.org/10.1007/s11356-023-25449-z

Ahmed, R. R., Streimikiene, D., Qadir, H. and Streimikis, J. (2023), "Effect of green marketing mix, green customer value, and attitude on green purchase intention: evidence from the USA", *Environmental Science and Pollution Research*, Vol. 30 No. 5, pp.11473–11495. doi: 10.1007/s11356-022-22944-7

Ahn, J. M., Koo, D. M. and Chang, H. S. (2012), "Different impacts of normative influences on pro-environmental purchasing behavior explained by differences in individual characteristics", *Journal of Global Scholars of Marketing Science*, Vol. 22 No. 2, pp.163–182. doi: 10.1080/12297119.2012.655098

Akehurst, G., Afonso, C. and Martins Gonçalves, H. (2012), "Re-examining green purchase behaviour and the green consumer profile: new evidences", *Management Decision*, Vol. 50 No. 5, pp.972–988, doi: 10.1108/00251741211227726

Al Zubaidi, N. (2020), "The relationship between collectivism and green product purchase intention: The role of attitude, subjective norms, and willingness to pay a premium", *Journal of Sustainable Marketing*, Vol. 1 No. 1, pp.24–32.

Al-Gasawneh, J. and Al-Adamat, A. (2020), "The mediating role of e-word of mouth on the relationship between content marketing and green purchase intention", *Management Science Letters*, Vol. 10 No. 8, pp.1701–1708. doi: 10.5267/j.msl.2020.1.010

Al-Quran, A. Z., Alhalalmeh, M. I., Eldahamsheh, M. M., Mohammad, A. A., Hijjawi, G. S., Almomani, H. M. and Al-Hawary, S. I. (2020), "Determinants of the green purchase intention in Jordan: the moderating effect of environmental concern", *International Journal of Supply Chain Management*, Vol. 9 No. 5, pp.366–371.

Alalei, A. and Jan, M. T. (2023), "Factors influencing the green purchase intention among consumers: an empirical study in Algeria", *Journal of Global Business Insights*, Vol. 8 No. 1, pp.49–65. doi: 10.5038/2640-6489.8.1.1181

Alamsyah, D. P., Udjaja, Y., Othman, N. A. and Ibrahim, N. R. W. (2020), "Green customer behavior: mediation model of green purchase", *International Journal of Psychosocial Rehabilitation*, Vol. 24 No. 05, pp.2568–2577. doi: 10.37200/IJPR/V24I5/PR201956

Albayrak, T., Aksoy, Ş. and Caber, M. (2013), "The effect of environmental concern and scepticism on green purchase behaviour", *Marketing Intelligence and Planning*, Vol. 31 No. 1, pp. 27–39. doi: 10.1108/02634501311292902

Ali, A., Khan, A. A., Ahmed, I. and Shahzad, W. (2011), "Determinants of Pakistani consumers' green purchase behavior: some insights from a developing country", *International Journal of Business and Social Science*, Vol. 2 No. 3, pp.217–226.

Ali, F., Ashfaq, M., Begum, S. and Ali, A. (2020), "How "Green" thinking and altruism translate into purchasing intentions for electronics products: the intrinsic-extrinsic motivation mechanism", *Sustainable Production and Consumption*, Vol. 24, pp.281–291. doi: 10.1016/j.spc.2020.07.013

Ali, M. (2021), "A social practice theory perspective on green marketing initiatives and green purchase behavior", *Cross Cultural and Strategic Management*, Vol. 28 No. 4, pp. 815–838. doi: 10.1108/CCSM-12-2020-0241

Alpert, M. I. and Peterson, R. A. (1972), "On the interpretation of canonical analysis", *Journal of Marketing Research*, Vol. 9 No. 2, pp.187–192. doi: 10.1177/002224377200900211

Ansari, M. Y. and Siddiqui, D. A. (2019), "Effects of culture on green purchase intention, the mediating role of new ecological paradigm, environmental collective efficacy and environmental knowledge", *Environmental Collective Efficacy and Environmental Knowledge* (December 29, 2019). doi: 10.5296/ijim.v5i1.16002

Arisbowo, N. and Ghazali, E. (2017), "Green purchase behaviours of Muslim consumers: An examination of religious value and environmental knowledge", *Journal of Organisational Studies and Innovation*, Vol. 4, pp.39–56

Ayodele, A. A., Panama, A. E. and Akemu, E. (2017), "Green awareness and consumer purchase intention of environmentally-friendly electrical products in Anambra, Nigeria", *Journal of Economics and Sustainable Development*, Vol. 8 No. 22. Available at SSRN: https://ssrn.com/abstract=3118901

Bailey, A. A., Mishra, A. and Tiamiyu, M. F. (2016), "GREEN consumption values and Indian consumers' response to marketing communications", *Journal of Consumer Marketing*, Vol. 33 No. 7, pp.562–573. doi: 10.1108/JCM-12-2015-1632

Baiquni, A. M. and Ishak, A. (2019), "The green purchase intention of Tupperware products: the role of green brand positioning", *Jurnal Siasat Bisnis*, Vol. 23 No. 1, pp.1–14. doi: 10.20885/jsb.vol23.iss1.art1

Bedard, S. A. N. and Tolmie, C. R. (2018), "Millennials' green consumption behaviour: exploring the role of social media", *Corporate Social Responsibility and Environmental Management*, Vol. 25 No. 6, pp.1388–1396. doi: 10.1002/csr.1654

Bhatt, R. and Bhatt, K. (2015), "Analyzing psychographic factors affecting green purchase intention", *Journal of Contemporary Research in Management*, Vol. 10 No. 1, pp.45.

Ch, T.R., Awan, T.M., Malik, H.A. and Fatima, T. (2021), "Unboxing the green box: an empirical assessment of buying behavior of green products", *World Journal of Entrepreneurship, Management and Sustainable Development*, Vol. 17 No. 4, pp. 690–710. doi: 10.1108/WJEMSD-12-2020-0169

Chaihanchanchai, P. and Anantachart, S. (2023), "Encouraging green product purchase: green value and environmental knowledge as moderators of attitude and behavior relationship", *Business Strategy and the Environment*, Vol. 32 No. 1, pp.289–303. doi: 10.1002/bse.3130

Chan, R. Y. (2001), "Determinants of Chinese consumers' green purchase behaviour", *Psychology and Marketing*, Vol. 18 No. 4, pp.389–413. doi: 10.1002/mar.1013

Chang, S. H. (2015), "The influence of green viral communications on green purchase intentions: the mediating role of consumers' susceptibility to interpersonal influences", *Sustainability*, Vol. 7 No. 5, pp.4829–4849. doi: 10.3390/su7054829

Chang, S.-H. and Chang, C.-W. (2017), "Tie strength, green expertise, and interpersonal influences on the purchase of organic food in an emerging market", *British Food Journal*, Vol. 119 No. 2, pp. 284–300. doi: 10.1108/BFJ-04-2016-0156

Chaudhary, R. and Bisai, S. (2018), "Factors influencing green purchase behavior of millennials in India", *Management of Environmental Quality*, Vol. 29 No. 5, pp. 798–812. doi: 10.1108/MEQ-02-2018-0023

Chen, C.C., Chen, C.W. and Tung, Y.C. (2018), "Exploring the consumer behavior of intention to purchase green products in belt and road countries: an empirical analysis", *Sustainability*, Vol. 10 No. 3, pp.854. doi: 10.3390/su10030854

Chen, L. (2013), "A study of green purchase intention comparing with collectivistic (Chinese) and individualistic (American) consumers in Shanghai, China", *Information Management and Business Review*, Vol. 5 No. 7, pp.342–346. doi: 10.22610/imbr.v5i7.1061

Chen, T. B. and Chai, L. T. (2010), "Attitude towards the environment and green products: consumers' perspective", *Management Science and Engineering*, Vol. 4 No. 2, pp.27–39.

Cheung, M. F. and To, W. M. (2019), "An extended model of value-attitude-behavior to explain Chinese consumers' green purchase behaviour", *Journal of Retailing and Consumer Services*, Vol. 50, pp.145–153. doi: 10.1016/j.jretconser.2019.04.006

Chi, N. T. K. (2021), "Understanding the effects of eco-label, eco-brand, and social media on green consumption intention in ecotourism destinations", *Journal of Cleaner Production*, Vol. 321, pp.128995. doi: 10.1016/j.jclepro.2021.128995

Dilotsotlhe, N. (2021), "Factors influencing the green purchase behaviour of millennials: an emerging country perspective", *Cogent Business and Management*, Vol. 8 No. 1, pp.1908745. doi: 10.1080/23311975.2021.1908745

Duong, C. D., Doan, X. H., Vu, D. M., Ha, N. T. and Dam, K. V. (2022), "The role of perceived environmental responsibility and environmental concern on shaping green purchase intention", *Vision*, pp.09722629221092117. doi: 10.1177/09722629221092117

Do Paço, A., Shiel, C., and Alves, H. (2019), "A new model for testing green consumer behaviour", *Journal of Cleaner Production*, Vol. 207, pp. 998–1006. 10.1016/j.jclepro.2018.10.105.

Eles, S. F. and Sihombing, S. O. (2017, June), "Predicting green purchase intention of generation: an empirical study in Indonesia", In *The 3rd PIABC (Parahyangan International Accounting and Business Conference*. Indonesia.

Erkan, I. and Evans, C. (2016), "The influence of eWOM in social media on consumers' purchase intentions: an extended approach to information adoption", *Computers in Human Behavior*, Vol. 61, pp.47–55. doi: 10.1016/j.chb.2016.03.003

Fabiola, K. and Mayangsari, L. (2020), "The influence of green skepticism, environmental knowledge and environmental concern on generation Z's Green purchase intentions in Indonesia", *Malaysian Journal of Social Sciences and Humanities (MJSSH)*, Vol. 5 No. 8, pp.96–105. doi: 10.47405/mjssh.v5i8.470

Fontes, E., Moreira, A. C. and Carlos, V. (2021), "The influence of ecological concern on green purchase behavior. Management & Marketing", *Challenges for the Knowledge Society*, Vol. 16 No. 3, pp.246–267. doi: 10.2478/mmcks-2021-0015

Ghazali, E. M., Mutum, D. S. and Ariswibowo, N. (2018), "Impact of religious values and habit on an extended green purchase behaviour model", *International Journal of Consumer Studies*, Vol. 42 No. 6, pp.639–654. doi: 10.1111/ijcs.12472

Goh, S. K. and Balaji, M. S. (2016), "Linking green skepticism to green purchase behaviour", *Journal of Cleaner Production*, Vol. 131, pp.629–638. doi: 10.1016/j.jclepro.2016.04.122

Green, P. E., Halbert, M. H. and Robinson, P. J. (1966), "Canonical analysis: an exposition and illustrative application", *Journal of Marketing Research*, Vol. 3 No. 1, pp.32–39. doi: 10.1177/002224376600300103

Grunert, S. C. and Juhl, H. J. (1995), "Values, environmental attitudes, and buying of organic foods", *Journal of Economic Psychology*, Vol. 16 No. 1, pp.39–62. doi: 10.1016/01 67-4870(94)00034-8

Gupta, M. and Syed, A.A. (2022), "Impact of online social media activities on marketing of green products", *International Journal of Organizational Analysis*, Vol. 30 No. 3, pp. 679–698. doi: 10.1108/IJOA-02-2020-2037

Hair, J.R., Anderson, R.E., Tatham, R.L. and Black, W.C. (1998), *Multivariate Data Analysis*, 5th ed.), Prentice Hall International Inc., NJ, USA.

Hamzah, M. I. and Tanwir, N. S. (2021), "Do pro-environmental factors lead to purchase intention of hybrid vehicles? The moderating effects of environmental knowledge", *Journal of Cleaner Production*, Vol. 279, pp.123643. doi: 10.1016/j.jclepro.2020. 123643

Han, H. (2020), "Theory of green purchase behavior (TGPB): a new theory for sustainable consumption of green hotel and green restaurant products", *Business Strategy and the Environment*, Vol. 29 No. 6, pp. 2815–2828. doi: 10.1002/bse.2545

Hariyanto, O. I. and Alamsyah, D. P. (2019), "The relationship of environmental knowledge and green purchase intention", *International Journal of Engineering and Advanced Technology (IJEAT)*. doi: 10.35940/ijeat.E1020.0585C19

Hasnah Hassan, S. (2014), "The role of Islamic values on green purchase intention", *Journal of Islamic Marketing*, Vol. 5 No. 3, pp. 379–395. doi: 10.1108/JIMA-11-2013-0080

Huang, Y.-C., Yang, M. and Wang, Y.-C. (2014), "Effects of green brand on green purchase intention", *Marketing Intelligence and Planning*, Vol. 32 No. 3, pp. 250–268. doi: 10.1108/MIP-10-2012-0105

Indriani, I. A. D., Rahayu, M. and Hadiwidjojo, D. (2019), "The influence of environmental knowledge on green purchase intention the role of attitude as mediating variable", *International Journal of Multicultural and Multireligious Understanding*, Vol. 6 No. 2, pp.627–635. doi: 10.18415/ijmmu.v6i2.706

Irandust, M. and Bamdad, N. (2014), "The role of customer's believability and attitude in green purchase intention", *Kuwait Chapter of the Arabian Journal of Business and Management Review*, Vol. 3 No. 7, pp.242.

Jain, V. K., Gupta, A., Tyagi, V. and Verma, H. (2020), "Social media and green consumption behavior of millennials", *Journal of Content, Community and Communication*, Vol. 10 No. 6, pp.221–230. doi: 10.31620/JCCC.06.20/16

Jaiswal, D. and Kant, R. (2018), "Green purchasing behaviour: a conceptual framework and empirical investigation of Indian consumers", *Journal of Retailing and Consumer Services*, Vol. 41, pp. 60–69. doi: 10.1016/j.jretconser.2017.11.008

Jaiswal, D. and Singh, B. (2018), "Toward sustainable consumption: investigating the determinants of green buying behaviour of Indian consumers", *Business Strategy and Development*, Vol. 1 No. 1, pp.64–73. doi: 10.1002/bsd2.12

Joshi, Y. and Rahman, Z. (2015), "Factors affecting green purchase behaviour and future research directions", *International Strategic Management Review*, Vol. 3 No. 1–2, pp.128–143. doi: 10.1016/j.ism.2015.04.001

Joshi, Y. and Rahman, Z. (2016), "Predictors of young consumer's green purchase behaviour", *Management of Environmental Quality*, Vol. 27 No. 4, pp. 452–472. doi: 10.1108/MEQ-05-2015-0091

Joshi, Y., Uniyal, D. P. and Sangroya, D. (2021), "Investigating consumers' green purchase intention: examining the role of economic value, emotional value and perceived marketplace influence", *Journal of Cleaner Production*, Vol. 328, pp.129638. doi: 10. 1016/j.jclepro.2021.129638

Kamalanon, P., Chen, J. S. and Le, T. T. Y. (2022), "Why do we buy green products?" An extended theory of the planned behavior model for green product purchase behavior", *Sustainability*, Vol. 14 No. 2, pp. 689. doi: 10.3390/su14020689

Kanchanapibul, M., Lacka, E., Wang, X. and Chan, H. K. (2014), "An empirical investigation of green purchase behaviour among the young generation", *Journal of Cleaner Production*, Vol. 66, pp.528–536. doi: 10.1016/j.jclepro.2013.10.062

Khaleeli, M., Oswal, N. and Sleem, H. (2021), "The moderating effect of price consciousness on the relationship between green products purchase intention and customers' purchase behavior: does environmental knowledge matters?", *Management Science Letters*, Vol. 11 No. 5, pp.1651–1658. doi: 10.5267/j.msl.2020.12.007

Khare, A. (2014), "Consumers' susceptibility to interpersonal influence as a determining factor of ecologically conscious behaviour", *Marketing Intelligence and Planning*, Vol. 32 No. 1, pp.2–20. doi: 10.1108/MIP-04-2013-0062

Khare, A., Mukerjee, S. and Goyal, T. (2013), "Social influence and green marketing: an exploratory study on Indian consumers", *Journal of Customer Behaviour*, Vol. 12 No. 4, pp.361–381. doi: 10.1362/147539213X13875568505903

Kusuma, P. N. P. D. and Handayani, R. B. (2018), "The effect of environmental knowledge, green advertising and environmental attitude toward green purchase intention", *Russian Journal of Agricultural and Socio-Economic Sciences*, Vol. 78 No. 6, pp.95–105.

Lee, K. (2009), "Gender differences in Hong Kong adolescent consumers' green purchasing behavior", *Journal of Consumer Marketing*, Vol. 26 No. 2, pp. 87–96. doi: 10.1108/07363760910940456

Lee, K. (2010), "The green purchase behavior of Hong Kong young consumers: the role of peer influence, local environmental involvement, and concrete environmental knowledge", *Journal of International Consumer Marketing*, Vol. 23 No. 1, pp.21–44. doi: 10.1080/08961530.2011.524575

Lee, K. (2014), "Predictors of sustainable consumption among young educated consumers in Hong Kong", *Journal of International Consumer Marketing*, Vol. 26 No. 3, pp.217–238. doi: 10.1080/08961530.2014.900249

Lee, Y. K. (2017), "A comparative study of green purchase intention between Korean and Chinese consumers: the moderating role of collectivism", *Sustainability*, Vol. 9 No. 10, pp. 1930. doi: 10.3390/su9101930

Lestari, E. R., Septifani, R. and Nisak, K. (2021, November), "Green awareness and green purchase intention: the moderating role of corporate image". In *IOP Conference Series: Earth and Environmental Science*, Vol. 924 No. 1, pp.012051. IOP Publishing. doi: 10.1088/1755-1315/924/1/012051

Li, H., Haq, I. U., Nadeem, H., Albasher, G., Alqatani, W., Nawaz, A. and Hameed, J. (2020), "How environmental awareness relates to green purchase intentions can affect brand evangelism? Altruism and environmental consciousness as mediators", *Revista Argentina de Clinica Psicologica*, Vol. 29 No. 5, pp.811–825. doi: 10.24205/03276716.2020.1079

Liao, Y. K., Wu, W. Y. and Pham, T. T. (2020), "Examining the moderating effects of green marketing and green psychological benefits on customers' green attitude, value and purchase intention", *Sustainability*, Vol. 12 No. 18, pp.7461. doi: 10.3390/su12187461

Liu, M.T., Liu, Y. and Mo, Z. (2020), "Moral norm is the key: an extension of the theory of planned behaviour (TPB) on Chinese consumers' green purchase intention", *Asia Pacific Journal of Marketing and Logistics*, Vol. 32 No. 8, pp. 1823–1841. doi: 10.1108/APJML-05-2019-0285

Luo, B., Sun, Y., Shen, J. and Xia, L. (2020), "How does green advertising skepticism on social media affect consumer intention to purchase green products?", *Journal of Consumer Behaviour*, Vol. 19 No. 4, pp.371–381. doi: 10.1002/cb.1818

Mahmoud, T. O., Ibrahim, S. B., Ali, A. H. and Bleady, A. (2017), "The influence of green marketing mix on purchase intention: the mediation role of environmental knowledge", *International Journal of Scientific and Engineering Research*, Vol. 8 No. 9, pp.1040–1048.

Mai, L. and Ness, M.R. (1999), "Canonical correlation analysis of customer satisfaction and future purchase of mail-order speciality food", *British Food Journal*, Vol. 101 No. 11, pp. 857–870. doi: 10.1108/00070709910301373

Malik, C., Singhal, N. and Tiwari, S. (2017), "Antecedents of consumer environmental attitude and intention to purchase green products: moderating role of perceived product necessity", *International Journal of Environmental Technology and Management*, Vol. 20 No. 5-6, pp.259–279. doi: 10.1504/IJETM.2017.091290

Mansoor, M. and Noor, U. (2019), "Determinants of green purchase intentions: positive word of mouth as moderator", *Journal of Business and Economics*, Vol. 11 No. 2, pp.143–160.

Martins, A. (2022), "Green marketing and perceived SME profitability: the meditating effect of green purchase behaviour", *Management of Environmental Quality*, Vol. 33 No. 2, pp. 281–299. doi: 10.1108/MEQ-04-2021-0074

Mas'od, A. and Chin, T. A. (2014), "Determining socio-demographic, psychographic and religiosity of green hotel consumer in Malaysia", *Procedia-social and Behavioral Sciences*, Vol. 130, pp.479–489. doi: 10.1016/j.sbspro.2014.04.056

Mei, O. J., Ling, K. C. and Piew, T. H. (2012), "The antecedents of green purchase intention among Malaysian consumers", *Asian Social Science*, Vol. 8 No. 13, pp.248. doi: 10.5539/ass.v8n13p248

Meloun, M. and Militký, J. (2011), "Statistical data analysis", *Statistical Analysis of Multivariate Data*, pp. 151–404. doi: 10.1533/9780857097200.151

Mishal, A., Dubey, R., Gupta, O.K. and Luo, Z. (2017), "Dynamics of environmental consciousness and green purchase behaviour: an empirical study", *International Journal of Climate Change Strategies and Management*, Vol. 9 No. 5, pp. 682–706. doi: 10.1108/IJCCSM-11-2016-0168

Mishra, P. (2015), *Business Research Methods*, Oxford University Press.

Moslehpour, M., Chau, K. Y., Du, L., Qiu, R., Lin, C. Y. and Batbayar, B. (2022), "Predictors of green purchase intention toward eco-innovation and green products: evidence from Taiwan". *Economic Research - Ekonomska Istraživanja*, pp.1–22. doi: 10.1080/1331677X.2022.2121934

Mostafa, M.M. (2006), "Antecedents of Egyptian consumers' green purchase intentions: a hierarchical multivariate regression model", *Journal of International Consumer Marketing*, Vol. 19 No. 2, pp.97–126. doi: 10.1300/J046v19n02_06

Mostafa, M. M. (2007), "Gender differences in Egyptian consumers' green purchase behaviour: the effects of environmental knowledge, concern and attitude", *International Journal of Consumer Studies*, Vol. 31 No. 3, pp.220–229.

Munamba, R. and Nuangjamnong, C. (2021), "The impact of green marketing mix and attitude towards the green purchase intention among generation Y consumers in Bangkok". Available at SSRN 3968444.

Naalchi Kashi, A. (2020), "Green purchase intention: a conceptual model of factors influencing green purchase of Iranian consumers", *Journal of Islamic Marketing*, Vol. 11 No. 6, pp. 1389–1403. doi: 10.1108/JIMA-06-2019-0120

Naz, F., Oláh, J., Vasile, D. and Magda, R. (2020), "Green purchase behavior of university students in Hungary: an empirical study", *Sustainability*, Vol. 12 No. 23, pp.10077. doi: 10.3390/su122310077

Nejati, M., Salamzadeh, Y. and Salamzadeh, A. (2011), "Ecological purchase behaviour: insights from a Middle Eastern country", *International Journal of Environment and Sustainable Development*, Vol. 10 No. 4, pp.417–432. doi: 10.1504/IJESD.2011.04 7774

Nekmahmud, M. and Fekete-Farkas, M. (2020), "Why not green marketing? Determinates of consumers' intention to green purchase decision in a new developing nation", *Sustainability*, Vol. 12 No. 19, pp.7880. doi: 10.3390/su12197880

Nekmahmud, M., Naz, F., Ramkissoon, H. and Fekete-Farkas, M. (2022), "Transforming consumers' intention to purchase green products: role of social media", *Technological Forecasting and Social Change*, Vol. 185, pp.122067. doi: 10.1016/j.techfore.2022.122067

Nguyen, T.N., Lobo, A. and Greenland, S. (2017), "The influence of cultural values on green purchase behaviour", *Marketing Intelligence and Planning*, Vol. 35 No. 3, pp. 377–396. doi: 10.1108/MIP-08-2016-0131

Nguyen, T. N., Lobo, A. and Nguyen, B. K. (2018), "Young consumers' green purchase behaviour in an emerging market", *Journal of Strategic Marketing*, Vol. 26 No. 7, pp.583–600. doi: 10.1080/0965254X.2017.1318946

Nia, B. P., Dyah, I. R., Hery, S. and Bayu, D. S. (2018), "The effect of green purchase intention factors on the environmentally friendly detergent product (Lerak)", In *E3S Web of Conferences*, Vol. 73, pp.06007. EDP Sciences. doi: 10.1051/e3sconf/20187306007

Noor, M. N. M., Jumain, R. S. A., Yusof, A., Ahmat, M. A. H. and Kamaruzaman, I. F. (2017), "Determinants of generation Z green purchase decision: a SEM-PLS approach", *International Journal of Advanced and Applied Sciences*, Vol. 4 No. 11, pp.143–147. doi: 10.21833/ijaas.2017.011.023

Noor, N. A. M., Muhammad, A., Kassim, A., Jamil, C. Z. M., Mat, N., Mat, N. and Salleh, H. S. (2012), "Creating green consumers: how environmental knowledge and environmental attitude lead to green purchase behaviour?", *International Journal of Arts and Sciences*, Vol. 5 No. 1, pp.55.

Ogiemwonyi, O., Harun, A. B., Alam, M. N., and Othman, B. A. (2020a). Do we care about going green? Measuring the effect of green environmental awareness, green product value and environmental attitude on green culture. An insight from Nigeria. *Environmental and Climate Technologies*, Vol. 24 No.1, 254–274. 10.2478/rtuect-2 020-0015

Ogiemwonyi, O., Harun, A.B., Alam, M.N., Karim, A.M., Tabash, M.I., Hossain, M.I., Aziz, S., Abbasi, B.A. and Ojuolape, M.A. (2020b), "Green product as a means of expressing green behaviour: A cross-cultural empirical evidence from Malaysia and Nigeria", *Environmental Technology and Innovation*, Vol. 20, p.101055. doi: 10.1016/j.eti.2020. 101055

Ok Park, J. and Sohn, S. H. (2018), "The role of knowledge in forming attitudes and behavior toward green purchase", *Social Behavior and Personality: An International Journal*, Vol. 46 No. 12, pp.1937–1953. doi: 10.2224/sbp.7329

Padel, S. and Foster, C. (2005), "Exploring the gap between attitudes and behaviour: understanding why consumers buy or do not buy organic food", *British Food Journal*, Vol. 107 No. 8, pp. 606–625. doi: 10.1108/00070700510611002

Panda, T. K., Kumar, A., Jakhar, S., Luthra, S., Garza-Reyes, J. A., Kazancoglu, I. and Nayak, S. S. (2020), "Social and environmental sustainability model on consumers' altruism, green purchase intention, green brand loyalty and evangelism", *Journal of Cleaner Production*, Vol. 243, pp.118575. doi: 10.1016/j.jclepro.2019.118575

Pop, R. A., Săplăcan, Z. and Alt, M. A. (2020), "Social media goes green—The impact of social media on green cosmetics purchase motivation and intention", *Information*, Vol. 11 No. 9, pp.447. doi: 10.3390/info11090447

Qomariah, A. and Prabawani, B. (2020, March), "The effects of environmental knowledge, environmental concern, and green brand image on green purchase intention with perceived product price and quality as the moderating variable", In *IOP Conference Series: Earth and Environmental Science*, Vol. 448 No. 1, pp.012115. IOP Publishing. doi: 10.1088/1755-1315/448/1/012115

Qureshi, M.A., Khaskheli, A., Qureshi, J.A., Raza, S.A. and Khan, K.A. (2022), "Factors influencing green purchase behavior among millennials: the moderating role of religious values", *Journal of Islamic Marketing*. doi: 10.1108/JIMA-06-2020-0174

Rahimah, A., Khalil, S., Cheng, J. M. S., Tran, M. D. and Panwar, V. (2018), "Understanding green purchase behavior through death anxiety and individual social responsibility: mastery as a moderator", *Journal of Consumer Behaviour*, Vol. 17 No. 5, pp.477–490. doi: 10.1002/cb.1733

Rahmi, D. Y., Rozalia, Y., Chan, D. N., Anira, Q. and Lita, R. P. (2017), "Green brand image relation model, green awareness, green advertisement, and ecological knowledge as competitive advantage in improving green purchase intention and green purchase behavior on creative industry products", *Journal of Economics, Business, and Accountancy Ventura*, Vol. 20 No. 2, pp.177–186. doi: 10.14414/jebav.v20i2.1126

Rizwan, M., Asif, R. M., Hussain, S., Asghar, M., Hassan, M. and Javeed, U. (2013), "Future of green products in Pakistan: an empirical study about green purchase intentions", *Asian Journal of Empirical Research*, Vol. 3 No. 2, pp.191–207.

Rizwan, M., Mahmood, U., Siddiqui, H. and Tahir, A. (2014), "An empirical study about green purchase intentions", *Journal of Sociological Research*, Vol. 5 No. 1, pp.290–305. doi: 10.5296/ jsr.v5i1.6567

Roberts, J.A. (1996), "Green consumers in the 1990s: profile and implications for advertising", *Journal of Business Research*, Vol. 36 No. 3, pp.217–231. doi: 10.101 6/0148-2963(95)00150-6

Ruangkanjanases, A., You, J. J., Chien, S. W., Ma, Y., Chen, S. C. and Chao, L. C. (2020), "Elucidating the effect of antecedents on consumers' green purchase intention: an extension of the theory of planned behaviour", *Frontiers in Psychology*, Vol. 11, pp.1433. doi: 10.3389/fpsyg.2020.01433

Samarasinghe, G. D. and Samarasinghe, D. S. R. (2013), "Green decisions: consumers' environmental beliefs and green purchasing behaviour in Sri Lankan context", *International Journal of Innovation and Sustainable Development*, Vol. 7 No. 2, pp.172–184. doi: 10.1504/IJISD.2013.053336

Sethi, V., Tandon, M. S. and Dutta, K. (2018), "A path model of antecedents of green purchase behaviour among Indian consumers", *International Journal of Public Sector Performance Management*, Vol. 4 No. 1, pp.21–44. doi: 10.1504/IJPSPM.2018. 088694

Sh. Ahmad, F., Rosli, N.T. and Quoquab, F. (2022), "Environmental quality awareness, green trust, green self-efficacy and environmental attitude in influencing green purchase behaviour", *International Journal of Ethics and Systems*, Vol. 38 No. 1, pp. 68–90. doi: 10.1108/IJOES-05-2020-0072

Sharma, A. and Foropon, C. (2019), "Green product attributes and green purchase behavior: a theory of planned behavior perspective with implications for circular economy", *Management Decision*, Vol. 57 No. 4, pp. 1018–1042. doi: 10.1108/MD-10-2018-1092

Sharma, A., Sharma, R. and Rana, G. (2022). Green organizational culture at the workplace of environmental sustainability. *In Digital disruption and Environmental, Social and Governance*. pp.124–137. Bazooka, New Delhi, India. ISBN 9789391363123.

Siddique, M.Z.R., Saha, G. and Kasem, A.R. (2021), "Estimating green purchase behavior: an empirical study using integrated behavior model in Bangladesh", *Journal of Asia Business Studies*, Vol. 15 No. 2, pp. 319–344. doi: 10.1108/JABS-04-2019-0120

Siyal, S., Ahmed, M. J., Ahmad, R., Khan, B. S. and Xin, C. (2021), "Factors influencing green purchase intention: moderating role of green brand knowledge", *International Journal of Environmental Research and Public Health*, Vol. 18 No. 20, pp.10762. doi: 10.3390/ijerph182010762

Sousa, S., Correia, E., Viseu, C. and Larguinho, M. (2022), "Analysing the influence of companies' green communication in college students' green purchase behaviour: an application of the extended theory of planned behaviour model", *Administrative Sciences*, Vol. 12 No. 3, pp.80. doi: 10.3390/admsci12030080

Sreen, N., Purbey, S. and Sadarangani, P. (2018), "Impact of culture, behavior and gender on green purchase intention", *Journal of Retailing and Consumer Services*, Vol. 41, pp.177–189. doi: 10.1016/j.jretconser.2017.12.002

Stern, P.C., Dietz, T. and Kalof, L. (1993), "Value orientations, gender, and environmental concern", *Environment and Behavior*, Vol. 25 No. 5, pp.322–348. doi: 10.1177/0013 916593255002

Stewart, D. and Love, W. (1968), "A general canonical correlation index", *Psychological Bulletin*, Vol. 70 No. 3, Pt.1, pp.160–163. doi: 10.1037/h0026143

Straughan, R.D. and Roberts, J.A. (1999), "Environmental segmentation alternatives: a look at green consumer behavior in the new millennium", *Journal of Consumer Marketing*, Vol. 16 No. 6, pp.558–575. doi: 10.1108/07363769910297506

Sun, Y., Leng, K. and Xiong, H. (2022), "Research on the influencing factors of consumers' green purchase behavior in the post-pandemic era", *Journal of Retailing and Consumer Services*, Vol. 69, pp.103118. doi: 10.1016/j.jretconser.2022.103118

Sun, Y. and Wang, S. (2019), "Understanding consumers' intentions to purchase green products in the social media marketing context", *Asia Pacific Journal of Marketing and Logistics*, Vol. 20 No. 4, pp.380–398. doi: 10.1108/APJML-03-2019-0178

Sun, Y. and Xing, J. (2022), "The impact of social media information sharing on the green purchase intention among generation Z", *Sustainability*, Vol. 14 No. 11, pp.6879. doi: 10.3390/su14116879

Tan, B. C. (2011), "The roles of knowledge, threat, and PCE on green purchase behaviour", *International Journal of Business and Management*, Vol. 6 No. 12, pp.14–27.

Tan, B. C. and Lau, T. C. (2011), "Green purchase behavior: examining the influence of green environmental attitude, perceived consumer effectiveness and specific green purchase attitude", *Australian Journal of Basic and Applied Sciences*, Vol. 5 No. 8, pp.559–567.

Taufique, K. M. R. and Islam, S. (2021), "Green marketing in emerging Asia: antecedents of green consumer behavior among younger millennials", *Journal of Asia Business Studies*, Vol. 15 No. 4, pp. 541–558. doi: 10.1108/JABS-03-2020-0094

Thi Tuyet Mai, N. (2019), "An investigation into the relationship between materialism and green purchase behavior in Vietnam and Taiwan", *Journal of Economics and Development*, Vol. 21 No. 2, pp. 247–258. doi: 10.1108/JED-10-2019-0044

Tudu, P. N. and Mishra, V. (2021), "To buy or not to buy green: the moderating role of price and availability of eco-friendly products on green purchase intention", *International Journal of Economics and Business Research*, Vol. 22 No. 2-3, pp.240–255. doi: 10.15 04/IJEBR.2021.116352

Uddin, S. F. and Khan, M. N. (2018), "Young consumer's green purchasing behavior: opportunities for green marketing", *Journal of Global Marketing*, Vol. 31 No. 4, pp.270–281. doi: 10.1080/08911762.2017.1407982

Veer, C., Kumar, P. and Sharma, R. (2022). Green Entrepreneurship: an avenue for innovative and sustainable product development and performance. In: *Entrepreneurial Innovations, Models, and Implementation Strategies for Industry 4.0*. CRC Press. USA. ISBN 9781032107936.

Verplanken, B. and Orbell, S. (2003), "Reflections on past behavior: a self-report index of habit strength 1", *Journal of Applied Social Psychology*, Vol. 33 No. 6, pp. 1313–1330. doi: 10.1111/j.1559-1816.2003.tb01951.x

Vu, D.M., Ha, N.T., Ngo, T.V.N., Pham, H.T. and Duong, C.D. (2022), "Environmental corporate social responsibility initiatives and green purchase intention: an application of the extended theory of planned behavior", *Social Responsibility Journal*, Vol. 18 No. 8, pp. 1627–1645. doi: 10.1108/SRJ-06-2021-0220

Wang, L., Wong, P. P. and Narayanan, E. A. (2020a), "The demographic impact of consumer green purchase intention toward green hotel selection in China", *Tourism and Hospitality Research*, Vol. 20 No. 2, pp.210–222. doi: 10.1177/1467358419848129

Wang, L., Wong, P. P. W. and Narayanan Alagas, E. (2020b), "Antecedents of green purchase behaviour: an examination of altruism and environmental knowledge", *International Journal of Culture, Tourism and Hospitality Research*, Vol. 14 No. 1, pp. 63–82. doi: 10.1108/IJCTHR-02-2019-0034

Wang, Y. M., Zaman, H. M. F. and Alvi, A. K. (2022), "Linkage of green brand positioning and green customer value with green purchase intention: the mediating and moderating role of attitude toward green brand and green trust", *Sage Open*, Vol. 12 No. 2, pp.21582440221102441. doi: 10.1177/21582440221102441

Xu, Y., Du, J., Khan, M. A. S., Jin, S., Altaf, M., Anwar, F. and Sharif, I. (2022), "Effects of subjective norms and environmental mechanism on green purchase behavior: an extended model of theory of planned behaviour", *Frontiers in Environmental Science*, Vol. 10, No. 39. doi: 10.3389/fenvs.2022.779629

Yadav, R. and Pathak, G. S. (2017), "Determinants of consumers' green purchase behavior in a developing nation: applying and extending the theory of planned behaviour", *Ecological economics*, Vol. 134, pp.114–122. doi: 10.1016/j.ecolecon.2016.12.019

Zafar, A. U., Shen, J., Shahzad, M. and Islam, T. (2021), "Relation of impulsive urges and sustainable purchase decisions in the personalized environment of social media", *Sustainable Production and Consumption*, Vol. 25, pp.591–603. doi: 10.1016/j.spc.2020.11.020

Zafar, S., Aziz, A. and Hainf, M. (2020), "Young consumer green purchase behaviour", *International Journal of Marketing Research Innovation*, Vol. 4 No. 1, pp.1–12. doi: 10.46281/ijmri.v4i1.493

Zahan, I., Chuanmin, S., Fayyaz, M. and Hafeez, M. (2020), "Green purchase behavior towards green housing: an investigation of Bangladeshi consumers", *Environmental Science and Pollution Research*, Vol. 27, pp.38745–38757. doi: 10.1007/s11356-020-09926-3

Zahid, M. M., Ali, B., Ahmad, M. S., Thurasamy, R. and Amin, N. (2018), "Factors affecting purchase intention and social media publicity of green products: the mediating role of concern for consequences", *Corporate Social Responsibility and Environmental Management*, Vol. 25 No. 3, pp.225–236. doi: 10.1002/csr.1450

Zhao, L., Lee, S.H. and Copeland, L.R. (2019), "Social media and Chinese consumers' environmentally sustainable apparel purchase intentions", *Asia Pacific Journal of Marketing and Logistics*, Vol. 31 No. 4, pp. 855–874. doi: 10.1108/APJML-08-2017-0183

2 E-Commerce and Green Packaging
Sustainable Business Trends

Selvi Kannan, Bhakti Parashar, and Amrita Chaurasia

2.1 INTRODUCTION

In 2021, global retail e-commerce sales reached a staggering 5.2 trillion US dollars. Projections indicate that this figure is poised to surge by 56% in the coming years, ultimately reaching an impressive 8.1 trillion dollars by 2026 (Chevalier, 2023). E-commerce, a subset of the broader concept of e-business, has emerged as a potent force in the world of commerce, serving as both a dynamic sales tool and a streamlined distribution system.

Unlike the traditional method of goods distribution, which relies heavily on intermediaries such as jobbers, wholesalers, and retailers, e-commerce operates as a direct distribution system. In essence, it leverages websites to gather product orders and facilitates the direct transfer of goods and services from manufacturers to end-users, effectively eliminating the need for intermediaries in the distribution process (Shankaraiah, 2018).

The sections below explore the remarkable growth of the e-commerce sector and its evolving recognition of the pivotal role played by consumers. Furthermore, these sections shed light on the emerging trends that focus on adopting sustainable packaging solutions. Businesses face an increasingly informed consumer base who are highly aware of the environmental and social consequences of their purchasing decisions.

Hence, in order to maintain their competitive edge, e-commerce businesses are consciously integrating socially responsible practices into their operations. This chapter will also showcase compelling success stories, highlighting the sustainable practices championed by environmentally responsible businesses. These stories serve as inspiring examples of the positive impact e-commerce can have when aligned with sustainability objectives.

DOI: 10.1201/9781003458944-2

2.1.1   E-COMMERCE

E-commerce, the practice of buying and selling goods or services online, has become an integral part of the modern business landscape, bringing about a transformation in how companies operate and engage with their customers (Morgan Stanley Euromonitor, 2022). Its initial surge was witnessed during the COVID-19 pandemic, as the need for online alternatives became paramount when physical stores closed, and people sought to stay safe indoors. The global e-commerce industry saw a significant boost, with its share of total retail sales rising from 15% in 2019 to 21% in 2021, and it currently stands at 22%, this growth trajectory indicates that the e-commerce sector, currently valued at $3.3 trillion, has the potential to reach an impressive $5.4 trillion by 2026 (Soft Landing, Hard Choices | Morgan Stanley, n.d., 2023) Notably, experts in equity research who focus on the American internet market, foresee e-commerce accounting for a remarkable 27% increase of all retail sales by 2026. There appears to be no definitive ceiling on the global penetration of e-commerce.

Global e-commerce nevertheless expanding rapidly and along comes the complex issue of wasteful packaging. The sector is estimated to generate over 100 million tons of packaging waste annually, posing a significant threat to the environment, including air and water pollution and contributing to climate change. The environmental impact of e-commerce has been on a steady rise (Escursell et al., 2021). For instance, Amazon, one of the prominent e-commerce giants, reported shipping over ten billion products in 2018, resulting in a carbon footprint of 44.40 million tons (Amazon, 2020; Escursell et al., 2020). Disturbingly, a recent analysis revealed that despite a 22% increase in net sales in 2019, Amazon's carbon emissions rose by 15% compared to the previous year (Amazon, 2020).

This pressing concern underscores the need for the e-commerce industry to address its environmental impact and take proactive measures to reduce packaging waste and lower its carbon footprint in order to sustain its growth in a responsible and eco-friendly manner.

2.1.2   ECO-FRIENDLY – GREEN PACKAGING

The pursuit of eco-friendly green packaging has been defined by a crucial criterion: not just recycling, but also providing customers with relevant information and the convenience of efficient waste handling (Palsson, 2018). With the escalating concern for environmental issues in recent years, consumers are increasingly demanding sustainable packaging solutions (Monnot et al., 2019). To address this demand, some companies have introduced innovative measures such as drones, electric vehicles, and convenient pickup points (Escursell et al., 2020; Mercedez-Benz, 2020).

The global reach of e-commerce platforms has expanded significantly, with more businesses engaging in cross-border trade to reach wider audiences. This expansion has been made possible through secure payment gateways, efficient logistics, and targeted marketing strategies (Beqiri, 2023). In Q1 2023 alone, total e-commerce sales saw a 12% year-on-year growth, with the B2B sector experiencing a 15% year-on-year increase, while the B2C sector exhibited slightly slower growth at 10% year-on-year. The B2B e-commerce market has reached a valuation of

$1.9 trillion, while the B2C market stands at $1.1 trillion (Beqiri, 2023). Both B2B and B2C e-commerce entities are placing sustainability at the forefront of their operations, emphasizing eco-friendly packaging materials, carbon emission reduction, and ethical sourcing practices. Projections suggest that the B2B industry may reach $2.1 trillion, and the B2C market could attain $1.3 trillion by the end of 2023.

Referred to by various terms like "eco-green packaging," "eco-friendly packaging," "sustainable packaging," or "recyclable packaging," this approach utilizes environmentally friendly materials while ensuring product efficiency and safety for both human and environmental well-being (Pauer, Wohner, Heinrich, & Tacker, 2019). The e-commerce sector is poised to maintain robust growth throughout 2023, driven by technological advancements, international trade expansion, and a growing commitment to sustainability.

The concept of "sustainable development" emerged around 1980, with the term gaining prominence in the Brundtland report (1987) Are (n.d.), which emphasized sustainable consumption in developed countries. Sustainable development rests on three fundamental pillars: social, economic, and environmental (Doordan, 2003; Jorgensen, 2013; Lyla and Seevakan, 2018; Manzini, 1994; Rizet et al., 2010; Shvarts, 2019; Rana & Arya 2023; Watson, 2001). Today, sustainable development encompasses three core areas: economic, environmental, and social. E-commerce businesses, like those in various industries, bear the responsibility of promoting responsible resource management and waste reduction.

Sustainable packaging comes in three forms, each tailored to the product and serving a specific purpose. Considerations include product packaging, materials, logistics optimization, freight costs, overall packaging sustainability, and post-consumption waste management (Bloom, 2023). Companies strive to meet external criteria by developing their own products with reduced environmental impact and greater use of recycled materials. For instance, Frosta introduced a novel container format made of 100% compostable paper instead of plastic, while other firms explored 100% plant-based plastic bottles. Finnish startup Sulapac specializes in creating biodegradable, moldable plastic alternatives for various applications, including cosmetics containers and drinking straws. Collaboration across the value chain is key to optimizing these solutions.

In 2023, researchers discovered over 170 trillion pieces of plastic in the world's waters, equivalent to more than 21,000 pieces for each of Earth's eight billion inhabitants (Bain & Company's Report, 2023). The total plastic production from 1950 to 2019 exceeded 9.5 billion tonnes, or more than one tonne per person alive today. Plastic waste management varies significantly by country and region, with high-income nations typically producing the most waste per person but having efficient waste management systems. Conversely, low- to middle-income nations struggle with inadequate waste management due to rapid industrialization and large coastal populations (OECD, 2023).

E-commerce packaging plays a critical role in conveying a brand's message to customers, protecting products during shipment, establishing brand identity, and managing shipping costs. In 1997, Pira International introduced new packaging criteria for the 21st century, while the European Union Packaging and Packaging Waste Directive (94/62/EU) implemented programs aimed at reducing packaging waste and promoting recycling (European Parliament, 1994; Sturges, 2000). By 2020, the EU set ambitious targets for recycling and recovery, aiming for 50% reuse and recovery of household waste and 70% reuse, recovery, and other forms of recovery (European Parliament, 2008).

Consumer behavior has also shifted, with 77% of shoppers in a 2014 study expressing greater environmental awareness and a willingness to pay extra for durable packaging (Hitchin and Bittermann, 2018; Kotler, 2017; Lindh, 2016). This underscores the growing importance of sustainable packaging solutions in e-commerce and the broader consumer landscape.

2.2   CIRCULAR ECONOMY

The concept of the circular economy extends far beyond recycling within the packaging industry. It encompasses every stage of a product's life cycle, from its initial design and production to distribution, use, and eventual recovery. This approach includes principles such as proximity and local integration.

Furthermore, it involves resource conservation, which includes:

1. Crafting products and packaging in an environmentally friendly manner.
2. Prudent utilization of resources.
3. Promoting the reuse of packaging, especially in business-to-business transactions.
4. Waste prevention by enhancing the recyclability of products and packaging.
5. The reuse of materials to optimize material flow.

In essence, the circular economy philosophy aims to create a sustainable and efficient ecosystem that minimizes waste and maximizes the utilization of resources throughout a product's entire life cycle.

Major developments of packaging in a circular economy are conceptualized in Figure 2.1 and explained in the following sections.

FIGURE 2.1   Development of packaging in a circular economy.

Source: Authors own.

2.2.1 BEGINNING OF MODERN PACKAGING

According to design historians, the beginnings of today's packaging go back to the years 1880-1900. During this period, new concepts such as mass marketing and mass production emerged, coinciding with the development of the first super-markets, with their low prices and large volumes, and the concept of strategic advertising on packaging (Twede, 2012). According to some authors (Piselli, 2016; Schleger, 1968), this revolution has influenced consumer behavior, stimulated innovative production automation techniques, supported the development of supermarkets and also brought economic benefits to packaging manufacturers and customers.

Folding cardboard boxes, tin, and glass were the primary packaging materials used from 1880 to 1900, and their industrialization allowed for the production of a large number of containers (Piselli, 2016; Twede, 2012).

2.2.2 DURING SECOND WORLD WAR

As materials were scarce during the Second World War (1939-1945), they were largely used to produce things for soldiers fighting overseas. The general populace was left with insufficient resources. Historically, most packaging was made from steel (containers), sheet metal (cans), glass (bottles), paper (containers), kraft paper (bags), fabric (bags), or wood (boxes), according to Risch (2009), Schleger (1968) and Sheldon (1944).

2.2.3 ERA OF MARKETING AND COMMUNICATION

The 1950s saw a boom in packaging-based marketing and communication, but also ushered in a period of waste problems and shorter product lifecycles known as "planned obsolescence" (Jorgensen, 2013). At the time, James Pilditch referred to packaging as the "silent seller" because it was so important to the success of products that it was almost as important as the products themselves (Cheskin, 1957; Maffei and Schifferstein, 2017; Opie, 1989; Pilditch, 1973; Vilnai-Yavetz, 2013). Self-service in supermarkets has made it easier and faster for customers to stock up on supplies (Quinn, 2012). Due to marketing's and designers' emphasis on shape, color and size to keep the product as visible as possible on the shelves until the 1960s (Cheskin, 1957; Silayoi and Speece, 2007; Vilnai-Yavetz, 2013), packaging materials remained unchanged from the 1960s until the 1960s.

2.2.4 GOLDEN AGE

The 1960s and 1970s are considered the "golden age" of technological progress and space exploration. Thanks to technological advances, the packaging industry now has access to innovative materials such as steam and water-resistant adhesives, aluminum and plastics. Although plastic was discovered in the 17th century, it was not used in packaging until the 1960s and 1970s due to its high flammability and limited lifespan (Wudl, 2014). Due to its ease of forming into numerous shapes, durability, hygiene, flexibility and low cost, plastic was the material of choice for all types of packaging at the time. According to Opie (1989) and Risch (2009), the most common materials used to make water bottles and plastic bags were polyethylene and polyethylene terephthalate (PET).

2.2.5 INTRODUCTION OF NANOTECHNOLOGY

Fortunately, advances in science and technology have produced a more humanistic outlook. As a result, materials can now be examined more closely thanks to advances in nanotechnology. Nanotechnology, according to Leydecker (2008) and MartinGago et al. (2014), can be highly useful in understanding the very small-scale physical, chemical, and biological aspects of materials.

2.2.6 CONCEPT OF SUSTAINABILITY

The term "sustainable development" was coined around 1980. Understanding the limited amount of fossil resources available for material production, as well as the negative impact of greenhouse gas (GHG) emissions from transport and other human activities, is crucial for people to enjoy nature (Doordan, 2003; Jorgensen, 2013; Lyla and Seevakan, 2018; Manzini, 1994; Rizet et al., 2010; Shvarts, 2019; Watson, 2001).

2.2.7 RECYCLING

In 1997, Pira International created new packaging criteria for the 21st century. Previously, the European Union Packaging and Packaging Waste Directive (94/62/ EU) (European Parliament, 1994) (Sturges, 2000) implemented a number of programs aimed at reducing packaging waste and promoting recycling. It was planned that by 2001, 25.4% of all packaging would be recycled, with 50.6% of all packaging waste recovered.

The European Parliament added "new recycling and recovery targets to be achieved by 2020: 50% willingness to reuse and recover some household waste and other household waste and 70% willingness to reuse, recover and other recovery". "Construction and demolition waste" was included in the Directive in 2008. According to a 2014 study by Cone Communications, 77% of shoppers were more environmentally conscious and made smarter food choices. In fact, many buyers were willing to pay 10% more than the normal price for the durable container (Hitchin and Bittermann, 2018; Kotler, 2017; Lindh, 2016).

2.2.8 INTRODUCTION OF CIRCULAR ECONOMY PACKAGE

The European Commission amended the law and published new suggestions as part of the "Circular Economy Package," which all EU countries were required to implement by 2030. To replace plastic and materials generated from fossil fuels, the packaging industry must clearly conduct research into and manufacture developing technologies. One technique for providing evidence is to highlight current market opportunities, for example for edible products and cellulosic products (Janjarasskul and Krochta, 2010). Proteins, carbohydrates, lipids and resins are used to classify food ingredients.

For example, Notpla's Ooho is an algae-based product that replaces plastic bottles with edible water bubbles. Kombucha, made by fermenting tea sugar, is another potential food ingredient. MakeGrowLab, a biodesign firm, discovered that it is ideal for packaging because it generates no waste (Shvarts, 2019).

According to Carlin (2019), the increasing use of paper and cardboard packaging materials can help ensure the sustainability of e-commerce. Mirjam de Bruijn (2016) has put together a new "Twenty" package for this series. This approach requires less cardboard because it reduces the water content of personal care, cosmetics and household cleaning products by 80%. Indeed, eliminating the need for plastic bottles and lowering packing materials reduces shipping costs to the point that customer behavior may shift (Solanki, 2018). These represent some of the key advancements in packaging. The growing emphasis from both governments and consumers on environmental impact and sustainability is prompting packaging companies to reassess their responsibilities in the manufacturing process. Businesses adopting a circular economy business model prioritize societal benefits, decouple economic activity from the depletion of finite resources, and aim to eliminate waste from the system (Lazard, 2021). Notably, consumer goods contribute to more than 60% of packaging.

Consumers are becoming increasingly conscious of the sustainability implications of their purchases. While the environmental impact of packaging is not a new concern, recent years have witnessed a significant surge in media coverage and consumer awareness regarding this issue. Plastic waste and ocean pollution, as exemplified by documentaries like *Blue Planet 2* telecast by BBC in 2017, brought heightened awareness to the general public (Hunt, 2017). These issues of pollution started to see a rising concern and subsequently in 2019 the estimated nine million premature deaths due to pollution caused alarm bells globally (Fuller et al, 2022). Consumer-facing businesses are now under greater pressure than ever to engage with their customers.

Currently, packaging follows a linear process of use and disposal, resulting in substantial waste and pollution. While this approach is convenient for consumers who do not bear the immediate costs of this waste, it represents a highly inefficient use of resources. Transitioning towards more sustainable packaging entails considering the complete life cycle of both the product and its packaging, with the ultimate goal of designing an entire system that operates in a circular manner. Companies that embrace this approach position themselves to stay ahead of regulatory changes, enhance their growth prospects, and reduce their environmental footprint (Lazard, 2021).

2.3   DESIGNING A SUSTAINABLE BUSINESS MODEL THAT EMBRACES GREEN PACKAGING IN E-COMMERCE

As the e-commerce sector experiences rapid growth, the demand for packaging is on the rise. However, traditional packaging materials like paper and plastic can have detrimental effects on the environment. Packaging serves a dual purpose in e-commerce – it stimulates consumption and safeguards products during delivery. Therefore, it becomes imperative to identify and utilize environmentally friendly materials for packaging in online shopping. This not only ensures the protection and transportation of products but also enhances the connection between products and consumers. Consequently, the transformation of express packaging and its promotion becomes essential, leading to a shift from express delivery to green logistics (Russo & Comi, 2016).

Brand attachment, as described by Park et al. (2010), is a crucial metric that gauges the depth of a consumer's affection or emotional bond with a product. Maintaining consistent brand loyalty is becoming increasingly challenging, emphasizing the need to explore ways to bolster brand associations and attachment among consumers. This is where the significance of green packaging in e-commerce comes into play.

E-commerce companies have access to a diverse range of green packaging options that can significantly reduce the environmental impact of shipping. For instance, using recyclable cardboard boxes or biodegradable packing materials can be a sustainable choice. To further minimize waste, businesses can opt for packaging materials made from recyclable or biodegradable substances like paper or plant-derived plastics.

Another aspect of green packaging in e-commerce is the concept of right-sizing. Right-sizing involves using packaging that is appropriately sized for the item being shipped, thus eliminating the need for excessive packaging materials. This not only reduces waste but also lowers shipping costs and enhances overall efficiency.

Green packaging is crafted from sustainable resources, including recycled paper, cardboard, and bioplastics. Moreover, it is designed to be recyclable or compostable, ensuring that it can be disposed of without harming the environment. Below are outlined some critical processes for E-commerce Green Packaging:

- **Choosing Sustainable Materials:** Selecting eco-friendly materials such as recycled paper, cardboard, and bioplastics for packaging.
- **Recyclability and Composability:** Ensuring that the packaging can be recycled or composted, minimizing its environmental impact upon disposal.
- **Right-Sizing:** Opting for packaging that is appropriately sized for the products being shipped, reducing waste and enhancing cost-efficiency.
- **Reducing Environmental Footprint:** Utilizing packaging options like recyclable cardboard and biodegradable materials to lower the overall environmental impact of shipping.

Incorporating these processes into e-commerce practices not only supports sustainability but also fosters brand loyalty by aligning with consumers' growing environmental consciousness.

Some of them are given below as E-Commerce Green Packaging Critical Process in Figure 2.2.

FIGURE 2.2   E-Commerce green packaging critical process.

Source: Authors own.

2.3.1   E-Commerce Green Packaging Critical Process

A remarkable 81% of respondents in a global study conducted by Neilsen expressed the view that it is of utmost importance for businesses to actively engage in environmental improvement programs. Furthermore, an impressive 73% of respondents indicated their willingness to modify their consumption habits in order to reduce their environmental impact. This sentiment aligns with the findings of a 2018 survey, which revealed a notable surge in the sales of organic products across various fast-moving consumer goods stores in the United States. Specifically, products boasting labels like "grass-fed" and "free-range" witnessed growth rates of 24% and 22%, respectively. It's worth noting that clean-label items now constitute one-third of dollar sales in the US beauty care sector, underscoring the increasing demand for environmentally conscious products. Notably, this trend has translated into a 16% boost in the sales of environmentally friendly products within the American chocolate market (Neilsen, 2018).

The global e-commerce market has demonstrated remarkable growth, surpassing the $9 trillion mark in 2019. Projections indicate that from 2020 to 2027, this market is expected to expand at a significant rate of 14.7%, with an annual growth rate of 5% (Coelho et al., 2020). This rapid expansion can be attributed to the widespread adoption of the internet and mobile phones, which have become indispensable tools in the daily lives of billions worldwide. Silva and Plsson (2022) have played a pivotal role in driving this digital transformation. Moreover, with the advancement of 5 G technology, increasing consumer tech awareness, and the promise of faster internet connections, the growth trajectory of e-commerce businesses is set to accelerate further (Coelho et al., 2020).

2.4   SUCCESS STORIES OF E-COMMERCE AND GREEN PACKAGING

Consumers today display heightened sensitivity to environmental concerns, gravitating towards seeing how businesses align with social, environmental and governance values. This confirms consumer behavior as Inmar Intelligence's 2022 study reveals that a substantial 54% of respondents aim to prioritize sustainable purchases most or all of the time. To address this paramount concern, organizations are taking pivotal steps towards sustainability, starting with the elimination of plastic in packaging in favor of 100% recyclable corrugated cardboard. This proactive shift not only underscores the company's commitment to environmental stewardship but also reduces the carbon footprint of each delivery, ultimately resulting in cost and resource savings. As businesses navigate the contemporary landscape, it becomes increasingly evident that sustainability is a key driver for both aspiring professionals and consumers, fostering brand loyalty, garnering favorable publicity, creating new avenues for business growth, and responding to societal pressures (Intelligence, 2022).

Effecting meaningful change in the e-commerce industry's sustainability practices involves a multi-faceted approach. One pivotal avenue is a thorough assessment of packaging processes, with an emphasis on transforming sustainability initiatives. As consumers seek ways to make more eco-conscious purchasing decisions, it is imperative for businesses to raise awareness of sustainable buying practices. The survey on sustainable buying practices underscores the significant opportunities for merchants to promote sustainability and offer environmentally friendly solutions to a rapidly expanding audience (Intelligence, 2022).

Innovations in packaging are also on the horizon, particularly through custom-fit parcels that eliminate excess packaging materials. Companies are exploring investments in automated packaging systems that swiftly and accurately construct boxes tailored to the size and type of orders, thus enhancing sustainability efforts. These technologies not only expedite the packaging process but also result in a 50% reduction in transportation volume, leading to fewer truckloads and reduced CO_2 emissions (Intelligence, 2022).

Furthermore, exemplifying the commitment to sustainability, Plaine Products, founded by two sisters, De Abreu et al. (2017), offers all-natural cosmetics and personal care items in reusable containers on a subscription basis. Their product line, including face wash, moisturizer, body wash, hand wash, body lotion, and shampoo, is vegan, cruelty-free, and biodegradable, all packed in recycled aluminum bottles. Plaine Products embodies the principles of a circular economy, ensuring nothing goes to waste. Customers are encouraged to return empty bottles for refills, fostering a sustainable, zero-waste approach (Ali, 2023).

Calvin Klein, a renowned luxury brand, is on a journey towards achieving zero waste by 2030. They have already made significant strides by making 74% of their packaging recyclable and using thinner materials in clothing packaging, resulting in over 200 tons of plastic saved annually. Additionally, they participate in the How2Recycle initiative, providing clear instructions on packaging to educate consumers about proper recycling. The parent company of Calvin Klein, PVH, has pledged to make its packaging 100% sustainable and ethically sourced, further highlighting its commitment to reducing environmental impact (Dobos, E. 2023).

Patagonia, an industry leader in sustainability, drew inspiration from the concept of a circular economy introduced by the book *Cradle to Cradle* in the 1990s. They initiated the Common Threads Garment Recycling program, focusing on recycling Capilene® baselayers to reduce petroleum reliance. Their partner, Teijin, utilized advanced chemical recycling to transform discarded baselayers into new polyester, contributing to resource efficiency and reducing waste (Ram, 2021). Patagonia continues to pioneer sustainable practices, including a 50% reduction in shipping volume by creating appropriately sized packages and incorporating recycled materials into a significant portion of its clothing line. They also prioritize social and environmental responsibility throughout their supply chain, holding suppliers to rigorous industry standards (Ram, 2021).

So, green packaging encompasses a wide array of options, including recyclable, biodegradable, reusable, and minimalist packaging. With growing awareness and emphasis on sustainability, the demand for green packaging is poised to soar in the years ahead.

Thus, in e-commerce, green packaging can be applied in a number of ways, including the following as conceptualized in Figure 2.3.

Packaging engineering plays a pivotal role in shaping the entire product life cycle, from production and packaging to transportation, with the overarching aim of minimizing adverse impacts throughout the product's journey and reducing waste generation in the realm of express package manufacturing (Bowns et al., 2018). To achieve this, a critical consideration lies in optimizing the utilization of materials,

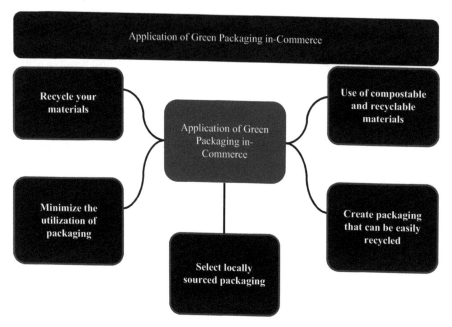

FIGURE 2.3   Application of green packaging in-Commerce.

Source: Authors own.

ensuring that they remain minimal while fulfilling the essential functions of packaging, including protection and transportation. Zhang (2022) asserts that it is imperative to approach packaging materials with careful consideration and wisdom, recognizing that their environmental impact extends from production to post-disposal, contingent on their utilization practices.

Intelligent packaging technology, as defined by Poelman et al. (2016), involves the incorporation of electrical components into either the packaging or the product itself, effectively highlighting the unique characteristics of the product. Green customer attitudes can assist businesses in developing more energy-efficient packaging. The foundational principles of packaging design are rooted in the concept of Green Design, which is dedicated to preserving the contemporary ecological environment. Smart packaging technology enables the tracking of product packaging data during the creation of low-carbon product packaging (Jiang et al., 2021). The advancements in intelligent packaging technologies exemplify the development of a scientific concept centered around a people-centric society. In today's society, individual personalities are given increasing consideration, and social progress acknowledges and accommodates everyone's individual development (Park & Taya, 2005). Smart packaging has witnessed significant growth since the turn of the century and is now among the world's top ten industries (He and Min, 2017).

The modern movement of goods relies increasingly on technological innovations within the packaging sector, driven by the rapid developments in business, science, and technology. This underscores how packaging design effectively achieves its multifaceted objectives, such as enhancing the aesthetic appeal of packaging, increasing the value of products, boosting profits, facilitating consumers' purchasing decisions, and conveying the cultural preferences of companies (Jeong et al., 2017).

The integration of artificial intelligence (AI) enhances product efficiency while also promoting market growth and environmentally friendly design (Sharma et al., 2021). Economic variables like income and food costs have a significant influence on individuals' dietary preferences. Families with limited incomes may face challenges in making healthier dietary choices due to the cost of food. The term "low-carbon" encompasses both a strategy and an indicator of carbon intensity, with the latter emphasizing a more conscientious approach to change. Lin et al. (2021) suggest that the increasing popularity of low-carbon lifestyles could be attributed to rising social, economic, and cultural norms.

The central focus of every project should be on the development of resource-efficient and environmentally friendly packaging. It's important to distinguish between "energy-saving packaging" and "environmentally packaged products," as these terms are often confused, leading to misconceptions about biodegradable packaging being inherently environmentally friendly. Packaging always serves as a reflection of the company's identity (Wang et al., 2019).

While AI is still a burgeoning technology, its potential to impact sustainable packaging is substantial. As AI technology advances, we can anticipate the emergence of more creative and efficient applications for AI in green packaging.

AI can, for example, monitor a customer's browsing patterns and suggest products of interest, potentially leading to overconsumption (Prakash et al., 2021), which can be detrimental to the environment. Additionally, the use of AI to create more immersive virtual experiences may reduce physical interactions with the outside world, potentially resulting in reduced physical activity and increased energy consumption.

Consequently, green packaging may come at a higher cost compared to traditional packaging, as finding and developing sustainable materials can be more challenging. Green packaging may also be heavier, potentially leading to increased shipping and transportation costs.

2.5   FURTHER RESEARCH

In identifying areas for further studies, it is critical to note that green packaging in e-commerce is emerging as an important focus for both practitioners and researchers. Firstly, in the chapter, the rise and importance of green packaging was discussed and this displays how consumer preferences align with e-commerce's success purely from a packaging acceptance point could be studied. It is clear from our chapter that consumers are concerned with environmental issues and e-commerce businesses are forced to adopt and adapt to eco-friendly packaging; however, how willing consumers are to take on the price difference and change their own behaviors is less understood. As the chapter has limitations, the study along with Smith (2022) also assessed the feasibility of utilizing materials derived from sources like algae, enhanced biodegradable polymers, and substances derived from agricultural waste. The adoption of these materials necessitates an evaluation of their cost-effectiveness and environmental impact when applied in real-world e-commerce scenarios can be analyzed.

Secondly, the quest to enhance sustainability can be further fortified by integrating AI and machine learning into packaging design and logistics. Deeper exploration can be conducted into automating sorting and recycling processes at distribution centers, optimizing packaging dimensions and types for specific products, and developing AI-powered predictive models for demand forecasting. These AI-driven technologies possess the potential to significantly reduce waste and resource consumption throughout the e-commerce supply chain.

Thirdly, forthcoming research should delve into consumer psychology, specifically scrutinizing consumer behaviors and preferences regarding environmentally friendly e-commerce packaging. By comprehending the factors that influence consumer choices, such as eco-labeling, convenience, and perceived value, businesses can craft packaging that resonates with environmentally conscious consumers.

Fourthly, the exploration of e-commerce packaging aligning with the principles of a circular economy presents an intriguing avenue for investigation. It is imperative to examine how businesses can design packaging that encourages product returns and reuse while concurrently upholding profitability and operational efficiency. The application of circular economy concepts has the potential to substantially curtail the generation of packaging waste and foster ethical corporate practices.

Finally, delving into the repercussions of regulatory changes on e-commerce enterprises and assessing strategies for compliance and competitive advantage forms another pertinent area of study. This multifaceted approach to research can contribute significantly to advancing sustainable practices in the e-commerce sector and mitigating its environmental footprint.

2.6 CONCLUSION

Green packaging without any doubt holds a pivotal role within the e-commerce industry, addressing pressing environmental challenges tied to packaging waste and carbon emissions. By embracing sustainable materials, right-sizing packaging, and implementing innovative designs, e-commerce enterprises can contribute significantly to a greener future while simultaneously meeting customer expectations. Amid the quest for valuing natural capital in economies seeking new resources, green marketing becomes a critical aspect in ensuring the sustainability of economic activities.

Scholars exploring the realms of business and consumer behavior are increasingly delving into the realm of green packaging alternatives. This underscores the growing relevance of climate change, environmental conservation, and resource preservation in contemporary society. Consumers, driven by concerns for the environment, factor various elements into their purchasing decisions regarding eco-friendly packaging. These factors include packaging design, material type, biodegradability, recyclability, and the country of origin. In summary, the fusion of green packaging with e-commerce stands at the forefront of sustainability endeavors within the modern corporate landscape. This study unit has provided a comprehensive examination of the current state of environmentally friendly materials, consumer-driven demands for green solutions, and waste reduction strategies in relation to sustainable packaging in e-commerce.

The future of e-commerce packaging is poised to undergo transformation through the integration of AI and innovation. Leveraging AI for enhancing logistics, refining packaging design, and automating recycling processes holds immense potential for reducing the industry's environmental footprint. It's essential to recognize that establishing sustainability in e-commerce packaging is an ongoing and evolving challenge.

Furthermore, companies actively engage in training their employees and indirectly influence labour policies within their supply chain partners. The scientific community invested in this field should adopt a more holistic approach to sustainable packaging, extending the focus beyond traditional research centered on packaging for food, beverages, and daily necessities to encompass a wide range of applications.

For start-up businesses, green marketing should be a consideration when defining their social responsibilities, adapting to evolving consumer preferences, and enhancing competitiveness. The risks associated with the progressive depletion of natural resources are also evident in areas such as eco-agritourist and natural product markets, with a strong emphasis on organic farming.

Ultimately, e-commerce businesses bear a responsibility to mitigate their environmental footprint, and one of the most crucial avenues for achieving this is through the adoption of green and sustainable packaging, with the support of AI technologies.

REFERENCES

Ali, S., & Shirazi, F. (2023). The paradigm of circular economy and an effective electronic waste management. Sustainability, 15(3), 1998.

Amazon. (2020). Carbon Footprint [online]. Retrieved from https://sustainability.aboutamazon.com/environment/sustainableoperations/carbon-footprint [Accessed 20 December 2020].

Are, F. O. F. S. D. (n.d.). 1987: Brundtland Report. https://www.are.admin.ch/are/en/home/media/publications/sustainable-development/brundtland-report.html products, P. (2023, September 11). sustainable-products. Retrieved from https://www.plaineproducts.com: https://www.plaineproducts.com/sustainable-products/

Bain & Company Report, P. &. (2023). Unpack the Power of Sustainable Packaging. Bain & Company's Global Forest Products: Brain and Company.

Beqiri, F. (2023, April 3). LinkedIn. Retrieved from www.linkedin.com: https://www.linkedin.com/pulse/global-e-commerce-market-summary-q1-2023-b2b-b2c-sectors-beqiri-1e/

Bowns, V., Jenkins, C., Njigha, N., Stratton, V., & Telfer-Wan, J. (2018). Growing issue of plastic marine debris – Addressing the growing issue of plastic marine debris within the central east coastline of Vancouver Island, British Columbia: Vancouver Island University.

Bloom. (2023, September 11). Retrieved from https://www.letsbloom.com: https://www.letsbloom.com/blog/eco-friendly-packaging/

Carlin, C. (2019). Tokyo pack introduces innovative plastic packaging ideas from Asian exhibitors. Plastics Engineering, 75(1), 8–13.

Cheskin, L. (1957). How to predict what people will buy. Liveright: Publishing Corporation, New York. OCLC Number : 245654572.

Chevalier, S. (2023). https://www.statista.com/statistics/379046/worldwide-retail-e-commerce-sales/. Retrieved from https://www.statista.com: https://www.statista.com/statistics/379046/worldwide-retail-e-commerce-sales/

Coelho, P., Corona, B., Klooster, R. T., & Worrell, E. (2020). Sustainability of reusable packaging–Current situation and trends. Resources Conservation & Recycling X, 6, 100037. 10.1016/j.rcrx.2020.100037

De Abreu, I. D., Hinojosa-Lindsey, M., & Asghar-Ali, A. A. (2017). A simulation exercise to raise learners' awareness of the physical and cognitive changes in older adults. Academic Psychiatry, 41(5), 684–687. 10.1007/s40596-017-0775-4.

De Bruijn, M., Amadou, A., Doksala, E. L., & Sangaré, B. (2016). Mobile pastoralists in Central and West Africa: Between conflict, mobile telephony and (im) mobility. Revue Scientifique et Technique-Office International des Epizooties, 35(2), 649–657.

Directive, H. A. T. (1994). European Parliament and Council Directive 94/36/EC of 30 June 1994 on colours for use in foodstuffs. Official Journal L, 237(10/09), 0013–0029.

Directive (2008/98/EC) on Waste (Waste Framework Directive) [WWW Document] Eur. Comm. https://ec.europa.eu/environment/waste/framework/index.htm

Dobos, E. (2023). From below or from above: How to force fashion MNCs to be more sustainable. Society and Economy, 45(3), 208–228.

Doordan, D.P. (2003). On materials. Design Issues, 19, 3e8. 10.1162/ 074793603322545000

Escursell, S., Massana, P. & Roncero, M. (2020). Sustainability in e-commerce packaging: A review. Journal of Cleaner Production, 280(1). 10.1016/j.jclepro.2020.12431.

Escursell, S., Llorach-Massana, P., & Roncero, M. B. (2021). Sustainability in e-commerce packaging: A review. Journal of Cleaner Production, 280, 124314. 10.1016/j.jclepro.2020.124314.

Fuller, R., Landrigan, P. J., Balakrishnan, K., Bathan, G., Bose-O'Reilly, S., Brauer, M., ... & Yan, C. (2022). Pollution and health: A progress update. The Lancet Planetary Health, 6(6), e535–e547.

He, Y., & Min, T. (2017). Design of indoor temperature monitoring and energy saving control technology based on wireless sensor. International Journal of Online Engineering, 13(7), 100.

Hitchin, T., & Bittermann, H. (2018). European consumer packaging perceptions study. Europe.

Hunt, E. (2017). Blue Planet II: From octopus v shark to fish that crawl, the series's biggest discoveries. The Guardian. Retrieved from https://www.theguardian.com/tv-and-radio/2017/nov/20/blue-planet-ii-what-have-we-learned-so-far.

Intelligence, I. (2022, July 19). Inmar Intelligence. Retrieved from Inmar.com: https://www.inmar.com/blog/press/new-inmar-intelligence-survey-reveals-divide-between-consumers-and-sustainability-e#:~:text=When%20making%20purchases%2C%2054%20percent,has%20more%20sustainable%20return%20practices.

Janjarasskul, T., & Krochta, J. M. (2010). Edible packaging materials. Annual Review of Food Science and Technology, 1, 415–448.

Jeong, J., Hong, T., Ji, C., et al. (2017). Development of a prediction model for the cost saving potentials in implementing the building energy efficiency rating certification. Applied Energy, 189(3), 257–270.

Jiang, N., Liu, H., Zou, J., et al. (2021). Packaging design for improving the uniformity of chip scale package (CSP). LED luminescence. Microelectronics Reliability, 122(2), Article 114136.

Jørgensen, F. A. (2013). Green citizenship at the recycling junction: Consumers and infrastructures for the recycling of packaging in twentieth-century Norway. Contemporary European History, 22(3), 499–516.

Kotler, P. (2017). Marketing 4.0: dal tradizionale al digitale. U. Hoepli – Torrossa. https://www.torrossa.com/it/resources/an/4149998.

Lazard. (2021, march 11). Packaging in the circular Economy. Retrieved from https://www.lazardassetmanagement.com: https://www.lazardassetmanagement.com/us/en_us/references/fundamental-focus/packaging-in-the-circular-economy.

Leydecker, S. (2008). Nano materials: In Architecture, Interior Architecture and Design. Walter de Gruyter.

Lin, C.Y., Chang, L.C., Chen, J.C., et al. (2021). Pain-administrable neuron electrode with wireless energy transmission: Architecture design and prototyping. Micromachines, 12(4), 356.

Lindh, H., Olsson, A., & Williams, H. (2016). Consumer perceptions of food packaging: Contributing to or counteracting environmentally sustainable development? Packaging Technology and Science, 29(1), 3–23.

Maffei, N. P., & Schifferstein, H. N. (2017). Perspectives on food packaging design. International Journal of Food Design, 2(2), 139–152.

Manzini, E. (1994). Design, environment and social quality: From" Existenzminimum" to" Quality Maximum". Design Issues, 10(1), 37–43.

Martín-Gago, J. Á., Llorente, C. B., Junquera, E. C., & Domingo, P. A. S. (2014). El Nanomundo en tus manos: las claves de la nanociencia y la nanotecnología. Grupo Planeta (GBS).

Mercedez-Benz. (2020). Vans and drones in Zurich [online]. Retrieved from https://www.mercedes-benz.com/en/vehicles/transporter/vans-drones-in-zurich/ [Accessed 20 December 2020].

Monnot, E., Reniou, F., Parguel, B. & Elgaaied-Gambier, L. (2019). "Thinking outside the packaging box": Should brands consider store shelf context when eliminating over-packaging? Journal of Business Ethics, 154, 355e370. 10.1007/s10551-017-34390.

Morgan Stanley Euromonitor, N. D. (2022). E-Commerce growth can stay stronger for longer. NY: https://www.morganstanley.com/ideas/global-ecommerce-growth-forecast-2022.morganstanley-Research. (14 June 2023). morganstanley. Retrieved from www.morganstanley.com: https://www.morganstanley.com/ideas/global-ecommerce- growth-forecast-2022

Neilson, J., Pritchard, B., Fold, N., & Dwiartama, A. (2018). Lead firms in the cocoa–chocolate global production network: An assessment of the deductive capabilities of GPN 2.0. Economic Geography, 94(4), 400–424.

OECD (2023), Global Plastics Outlook: Plastics use in 2019, OECD Environment Statistics (database). OECD Environment Statistics (database), 10.1787/efff24eb-en (accessed on 11 September 2023).

Opie, R. (1989). Packaging source book. (No Title).

Park, C. W. (2010). Brand attachment and brand attitude strength: Conceptual and empirical differentiation of two critical brand equity drivers. https://www.semanticscholar.org/paper/Brand-Attachment-and-Brand-Attitude-Strength%3A-and-Park-MacInnis/a597be324727b93bb86020d022287b8bfa7ee7f8.

Park, J. J., & Taya, M. (2005). Design of thermal interface material with high thermal conductivity and measurement apparatus. Journal of Electronic Packaging, 128(1), 46–52. 10.1115/1.2159008.

Pauer, E., Wohner, B., Heinrich, V., & Tacker, M. (2019). Assessing the environmental sustainability of food packaging: An extended life cycle assessment including packaging-related food losses and waste and circularity assessment. Sustainability, 11, 925.

Pilditch, J. (1973). The silent salesman: How to develop packaging that sells.

Piselli, A., Lorenzi, A., Alfieri, I., Garbagnoli, P., & Del Curto, B. (2016). Designing sustainable scenarios: Natural-based coatings as a barrier to oil and grease in food paper packaging. International Journal of Designed Objects, 10(1).

Poelman, M.P., Eyles, H., Dunford, E., et al. (2016). Package size and manufacturer-recommended serving size of sweet beverages: A cross-sectional study across four high-income countries. Public Health Nutrition, 19(6), 1008–1016.

Prakash, C., Saini, R., & Sharma, R. (2021). Role of Internet of Things (IoT) in Sustaining disruptive businesses. In R. Sharma, R. Saini, C. Prakash, & V. Prashad, Internet of Things and Businesses in a Disruptive Economy (1st ed.). New York: Nova Science Publishers.

Pålsson, H. (2018). Packaging Logistics: Understanding and Managing The Economic and Environmental Impacts of Packaging in Supply Chains. London: Kogan Page.

Quinn, S. F. (2012). Crowning the customer: How to become customer-driven. Ireland: The O'Brien Press.

Ram, A. (2021, March 10). Our Quest for Circularity. Retrieved from https://www.patagonia.com/social-responsibility/: https://www.patagonia.com/social-responsibility/.

Rana, G., and Arya, V. (2023), "Green human resource management and environmental performance: mediating role of green innovation – A study from an emerging country", Foresight. 10.1108/FS-04-2021-0094.

Risch, S. J. (2009). Food packaging history and innovations. Journal of Agricultural and Food Chemistry, 57(18), 8089–8092.

Rizet, C., Cornelis, E., Browne, M., & Leonardi, J., (2010). GHG emissions of supply chains from different retail systems in Europe. Procedia – Social and Behavioral Science, 2, 6154. 10.1016/j.sbspro.2010.04.027.

Russo, F., & Comi, A. (2016). Urban freight transport planning towards green goals: Synthetic environmental evidence from tested results. Sustainability, 8(4), 381. 10.3390/su804038.

Schleger, H. (1968). Package and print. The Development of Container and Label Design.

Seevakan, L. P. (2018). Advancement of cubosomes nanomedicine. International Journal of Pure and Applied Mathematics, 119(12), 7847–7854.

Shankaraiah, D. M. (2018). E-commerce growth in India: A study of segments contribution. Academy of Marketing Studies Journal, 22 (2).

Sharma, R., Saini, R., Prakash, C., & Prasad, V. (2021) Internet of Things and Businesses in a Disruptive Economy, 1st edn., New York: Nova Science Publishers.

Sheldon, C. L. (1944). Containers go to war. Harvard Business Reviews, 22, 220.

Shvarts, J (2019) Packaging Strategies. (n.d.-c). https://www.packagingstrategies.com/articles/94829-eco-friendly-packaging-materials-of-the-future.

Silayoi, P. & Speece, M., (2007). The importance of packaging attributes: a conjoint analysis approach. European Journal of Marketing, 41, 1495e1517. 10.1108/0309056071 0821279.

Silva, N., & Pålsson, H. (2022). Industrial packaging and its impact on sustainability and circular economy: A systematic literature review. Journal of Cleaner Production, 333, 130165. 10.1016/j.jclepro.2021.130165.

Smith, Z. A., & Jacques, P. (2022). The Environmental Policy Paradox. New York: Taylor & Francis.

Solanki, S. (2018). Why materials matter: Responsible design for a better world.

Soft landing, hard choices | Morgan Stanley. (n.d.). Morgan Stanley. https://www.morganstanley.com/ideas/investment-outlook-mid-year-2023-global-risk

Sturges, M. (2000). Environmental legislation for packaging in the next millennium. Packaging Technology and Science: An International Journal, 13(1), 37–42.

Twede, D. (2012). The birth of modern packaging: Cartons, cans and bottles. Journal of Historical Research in Marketing, 4(2), 245–272.

Vilnai-Yavetz, I., & Koren, R. (2013). Cutting through the clutter: purchase intentions as a function of packaging instrumentality, aesthetics, and symbolism. The International Review of Retail, Distribution and Consumer Research, 23(4), 394–417. 10.1080/095 93969.2013.792743.

Wang, C., Guo, X. & Zhu, Y. (2019). Energy saving with optic variable wall for stable air temperature control. Energy, 173(4), 38–47.

Watson, P., (2001). A Terrible Beauty. A History of the People and Ideas that Shaped the Modern Mind. Phoenix.

Wudl, F., (2014). The bright future of fabulous materials based on carbon. Daedalus, 143, 31

Zhang, S. (2022). Research on energy-saving packaging design based on artificial intelligence. Energy Reports, 8, 480–489. 10.1016/j.egyr.2022.05.069.

3 Synergy of Green Entrepreneurship Among Indigenous Community
Case for Sri Lanka

Imali Fernando, Hirusha Amarawansha, Sewwandika Gamage, Jami Perera, and Nipuni Fernando

3.1 INTRODUCTION

3.1.1 INDIGENOUS COMMUNITY IN THE CONTEXT OF SRI LANKA

Sri Lanka is an island bordering India's mainland, connected by the Palk Strait and the Gulf of Mannar. People from all over the world have settled in Sri Lanka over time. The country has a population of around 21 million people who speak different languages, come from different ethnic backgrounds, and practice different religions. The majority of Sri Lankans are Buddhist Sinhalese, accounting for around 74.9% of the total population. Moreover, Sri Lankan Tamils, who speak Tamil and practice Hinduism, account for around 15.3% of the total population. In addition, Muslims constitute 9.3% of the total population and some minorities account for less than 0.5% of the total population (Central Bank of Sri Lanka (CBSL), 2021). This minority group includes the Muslim Malays, Christians, and the Vedda people. Further, the "Vedda" community is believed to be the most indigenous group in Sri Lanka. Further, the language of the Vedda people is a mix of older Vedda and Sinhala. Most the people use term "Vadda" to identify the indigenous community in Sri Lanka and this term implies "the person who uses bows and arrows", which reflects their practices such as cultivation, hunting, and collecting forest products, etc. (Attanapola & Lund, 2013).

Indigenous peoples all over the world have a strong bond with their ancestral lands, cultural history, and traditional rituals. These villages in Sri Lanka are no exception, exemplifying a wonderful symbiotic relationship between their way of life and the natural environment. As the Vedda people the one of Sri Lanka's indigenous communities, their customs, norms, and subsistence tactics are directly

DOI: 10.1201/9781003458944-3

linked to the cycles of the forests in which they live (Dharmadasa, 1993). Furthermore, the Tamil-speaking indigenous populations, which are situated in the island's north and east have a rich tapestry of traditions that reflect their intimate relationship with the land and sea. As a result, they are practicing sustainability throughout their all activities and practices from generation to generation. Thus, the sustainable resource management techniques of those indigenous people clearly despite the bond they have with nature. Additionally, the traditional ecological knowledge passed down through generations enables them to live in harmony with the environment (Ford et al., 2020). This information includes insights into soil fertility, weather patterns, plant and animal behavior, and ways for sustainable harvesting. Hence, this indigenous wisdom has supported their green commercial pursuits, instilling a feeling of ecological responsibility in their economic actions.

The generational transmission of their knowledge is an important part of the indigenous worldview. Accordingly, elder generations are respected as wisdom keepers, and their teachings play an important role in forming the collective consciousness of those communities. As modern issues threaten their traditional way of life, indigenous groups are using their traditional wisdom to develop new solutions. Some of the tactics used by these communities to engage in green entrepreneurship include the conservation of endemic flora, the resurrection of traditional farming practices, and the development of eco-friendly tourism. However, recognition of the vital role that indigenous groups play in maintaining environmental balance has taken much attention in research agendas and practitioners around the world. As a result, this designation is an important step toward enabling these communities to use their green practices for long-term development. Today, there is a strong desire among consumers worldwide to safeguard the environment, and consumer behavior is shifting toward environmentally friendly or green products (Sharma & Kushwaha, 2015). As a result, entrepreneurs introduce green products and technologies to the market and lay the groundwork for green entrepreneurship by transforming patterns and prototypes into tangible commercial products, and green entrepreneurs eventually introduce green products and technologies to the market (Nikolaoua, Ierapetritis, & Tsagarakis, 2011) As a result, the incorporation of these indigenous viewpoints with modern principles of green entrepreneurship provides prospective options for long-term economic success while protecting their cultural identity and natural surroundings.

Even though, green entrepreneurship has received much attention for its environmental and economic ramifications, its application within the Sri Lankan context and indigenous communities has rarely been studied yet. Thus, this study focuses on discovering how the concepts of ecologically sustainable business operations match the cultural values, traditions, and economic goals of an indigenous community in Sri Lanka. Drawing from the above gap, the objective of this study is to explore green entrepreneurship among the Indigenous Community in Sri Lanka with a focus on their role and contribution to sustainability.

Indigenous groups of Sri Lanka are spread over 19 different areas (Figure 3.1) Among these, three locations stand out due to their significantly higher populations in comparison to the rest and those three locations include *Dabana*, *Wakare,* and *Rathugala*. Moreover, indigenous groups in these locations frequently have distinct

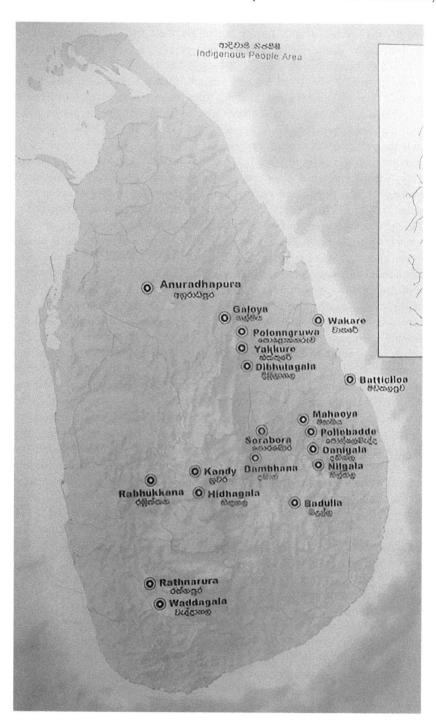

FIGURE 3.1   Map of Indigenous community locations in Sri Lanka.

Source: Central Cultural Fund.

cultural practices, traditions, and ways of life from the general population and they contribute immensely to the great diversity of the cultural landscape of Sri Lanka.

Dabana has the most indigenous people of the three locations and it is recognized as the most notable site for indigenous communities in Sri Lanka. The Dabana community is well-known for its magnitude in comparison to other indigenous communities in Sri Lanka. They retain their own distinct customs and beliefs, which are founded in their origin. Rathugala comes in second place after Dabana due to its distinct indigenous community. While the Rathugala community has certain similarities with the Dabana, they have their own set of traditions and beliefs that define them. Wakare is third in terms of attention, with a smaller indigenous population than Dabana and Rathugala. Further, Wakare is also unique in its indigenous Tamil inhabitants engage in fishing as their main source of income. Accordingly, this study selected two areas including Dabana and Rathugala by considering the population, distinct customs, and beliefs of this group.

3.1.2 Green Entrepreneurial Practices and Sustainability

In Sri Lanka, indigenous groups have embraced a variety of entrepreneurial practices that are closely related to the environment and are significantly informed by their traditional ecological knowledge. These practices cover across industries and demonstrate an in-depth understanding of how they balance economic activities and environmental well-being. Accordingly, Indigenous communities are using mixed planting practices cleverly, incorporating fruit-bearing trees, medicinal plants, and agricultural crops. However, as per the observations and discussions, the indigenous community in Dabana engages more with agricultural activities compared to people in Rathugala. This not only diversifies their revenue streams, but also enhances soil fertility, biodiversity, and water conservation as well. Furthermore, the communities' deep respect for their natural resources is visible in their rigorous harvesting procedures, which ensure a lower negative environmental impact. Indigenous people in Sri Lanka practice robust food systems by reusing old seeds and growing local foods while working in groups and helping each other with farming. These systems exhibit a thorough understanding of ecosystem services and a holistic approach to sustainability.

Further, those indigenous people in Sri Lanka also utilize their business spirit through sustainable crafts and artistry as shown in Figure 3.2. Thus, they are engaging in traditional weaving, ceramics, and handicrafts and those practices clearly demonstrate their artistic abilities while using locally available, biodegradable materials. Moreover, these items have cultural importance, emphasizing the link between indigenous identity and long-term perspective.

Sri Lankan indigenous people have a special bond with the forest. They travel into the forest in search of herbs, bee honey, etc. After collecting forest products, these indigenous groups bring their forest findings to town and sell them in local stores. This activity not only demonstrates their intimate knowledge of nature's gifts but also allows them to share the benefits of these natural resources with a larger community as well. These indigenous communities bridge the gap between traditional wisdom and contemporary requirements by connecting the forest and urban areas, all while preserving their cultural legacy.

FIGURE 3.2 Traditional biodegradable crafts by the Indigenous community.

In addition, indigenous people in those areas are practicing eco-tourism to generate sustainable income. They raise environmental awareness by sharing their extensive ecological expertise and cultural legacy with tourists. Those activities not only diversify income sources but also promote the local economy, reinforcing the link between environmental health and economic well-being. These communities have embraced ecotourism as a means of sharing their culture, heritage, and natural surroundings with visitors in a sustainable manner. Accordingly, those unique activities include showcasing traditional dance and performances, culinary experience, guiding them for nature walks, and traditional storytelling, etc.

Green entrepreneurship activities in indigenous communities in Sri Lanka are important for more than just economic reasons, they also contribute to community empowerment and sustainability. These activities strengthen the indigenous way of life by connecting their operations with traditional knowledge and values and ultimately contributing to the preservation of both culture and ecology.

3.1.3 Objectives of the Study

- Exploring the green entrepreneurship practices among the Indigenous community in Sri Lanka with a focus on their role and contribution to sustainability
- Exploring the innovative green business models and entrepreneurial challenges faced by the Indigenous community in Sri Lanka
- Synthesis of the strategies adopted by the Indigenous community in Sri Lanka to overcome green entrepreneurial challenges

3.2 LITERATURE REVIEW

3.2.1 Indigenous Knowledge

Indigenous knowledge is a traditional knowledge embedded within the traditional communities within the countries. According to the Senanayake (2006) explanation

Indigenous knowledge is the result of practical participation in daily life, and it continually gets strengthened by expertise and experimentation. Indigenous knowledge systems have a broad understanding of ecosystems and sustainable resource use. Today, there is a grave risk of losing much indigenous knowledge, as well as valuable knowledge about how to live sustainably both ecologically and socially. Moreover, he explains that Indigenous knowledge is regarded social capital of the people who are in socially deprived societies. It is their primary strength to invest in the struggle for their existence within the society for example producing food, providing shelter, and gaining control over their personal human lives. Furthermore, Senanayaka (2006) explains that Indigenous knowledge is specific to a region. It originates in a specific location with a series of experiences, and it is created by people who live in those locations. As a result, transmitting indigenous knowledge to other locations may result in the destruction of location-based advantage in knowledge creation.

Concerning the Indigenous community, Padmasiri (2018) explains in their study that Communities in Sri Lanka have an amazing reservoir of insufficiently utilized, neglected, and undetected indigenous knowledge and skills, which have mostly become ineffective or disappeared due to a variety of factors. The indigenous knowledge system, particularly in medicine, is mostly unspoken. Controlling indigenous knowledge of indigenous medicine is thus a significant challenge. Economic autonomy as well as sustainability are thought to be achievable through a hybrid system of development that combines existing Indigenous knowledge and modern technologies. The management of indigenous knowledge will update perish traditions and encourage community-based participation in a country's development programs.

3.2.2 INDIGENOUS COMMUNITY IN SRI LANKA

According to the World Bank (2019), The world's indigenous peoples number nearly 370 million, accounting for roughly 5% of the global population. The vast majority (260 million, or 70%) live in Asia-Pacific countries. Moreover, according to the World Bank's explanations, they already face numerous challenges as a result of both economic and political discrimination, loss of land and resources, violations of human rights, prejudice, food insecurity, and an absence of financial possibilities. The word Indigenous has been commonly employed for a long time, regardless of the fact that the popularity of other terms, for example, Indian, native, local, and First Nations people groups, can vary depending on the geological context. Moreover, he explains that Indigenous people are referred to in Sri Lanka as, "Adivasi" or as "Vedda". According to another study done by Ranasinghe and Cheng in 2017, Several indigenous communities retain important intangibles rooted in their cultures, including choreography, narratives, hunting techniques, melodies, and their language its own. Moreover, Indigenous communities are detected as independent cultural communities with ancestral homelands or geographically distinct traditional ecosystems. They are the descendants of the original groups who settled in the area prior to the formation of modern states and borders. They maintain their cultural and social distinctiveness by distancing themselves from the dominant society or culture (Galappaththi et al. 2020).

According to De Silva and Punchihewa (2011), The "Vedda" are the ancestors of the island's initial neolithic society as a whole who have resided there since the 6th century BC (Lund and Attanapola, 2013). Vedda's population was reported to be 0.0044% of Sri Lanka's total population in 2011. Moreover, Galappaththi et al. (2020) explain the nature of the current Adivasi community and explain that there has been no recent census to figure out the precise number of people living in the community of Vedda. Various Vedda groups live on the island, including Dambana-Vedda, Pollebadda-Vedda, Rathugala-Vedda, and Coastal-Vedda. Vedda's choices for livelihood vary depending on where they live. For food security and nutrition, many Veddas are currently dependent on subsistence and commercial livelihood activities. Many of them engage in paddy farming and other agriculture, also known as "chena" in Sri Lanka. Some people collect honey from bees, yams, and fruits. Vedda are coastal people who rely extensively on fisheries-related tasks such as culture-based fisheries.

3.2.3 INDIGENOUS ENTREPRENEURSHIP

Indigenous entrepreneurship has been described as "the development, management, and creation of new ventures by Indigenous peoples for the benefit of Indigenous peoples" (Lansdown, 2005). Indigenous populations live frequently close to the environment and rely on easily accessible assets such as animals or fish. Indigenous business is frequently environmentally friendly. Much of the economic entrepreneurial activity carried out by indigenous people doesn't take place in modern-day markets. In certain instances, activities take place in spite of any kind of exchange market; these are known as internal subsistence activities (Dana, 2015). Moreover, according to Dana (2015), the informal sector is important among indigenous peoples and they are engaging in small-level entrepreneurial activities. While large-scale entrepreneurship, requires an extensive infrastructure, for communications, information, transportation, and capital. Indigenous entrepreneurship is usually ecologically sustainable; this frequently allows people of Indigenous descent to depend on easily accessible resources, and as a result, work in Indigenous communities is frequently irregular. Indigenous peoples' social organization is frequently based on familial relationships and does not always evolve due to market needs (Dana, 2015).

3.2.4 INDIGENOUS COMMUNITY AND GREEN ENTREPRENEURSHIP

According to Tien et al. (2023) explanation based on the World Commission on Environment and Development, they identify ecological sustainability or sustainable development as growth that caters to current needs without jeopardizing the capacity of future generations to satisfy their own requirements. Similarly, to the century-old notion of social responsibility, the recently emerged idea of sustainability or sustainable development has been thoroughly examined in many academic investigations. Moreover, they suggest that the concept of sustainability or sustainable development becomes increasingly prevalent today since it establishes widespread patterns in all areas of business activity, as well as in all fields of modern research and development, academic as well as practical. The Triple Bottom Line (TBL) model emphasizes the three critical dimensions of firm performance: the economy, society, and the environment (Asadi et al., 2020). The triple bottom line concept clearly underlines the way to deal with green entrepreneurship.

The surge in green entrepreneurship has paralleled the escalating focus on environmental safeguarding from diverse stakeholders. The core of green entrepreneurship lies in green innovation, a focal point that has garnered substantial scholarly interest. Various terms, including environmental entrepreneurship, have been employed to define and summarize the concept of green entrepreneurship (Veer et al., 2022). Moreover, Yin et al. (2022) cited Lee & Min (2015) and explained that Green entrepreneurship refers to the entrepreneurial actions undertaken within an organization to innovate products, processes, and other aspects with the dual aim of generating profit while prioritizing environmental protection (Sharma & Rana 2021). At its core, green entrepreneurship involves the development and implementation of novel ideas that contribute to sustainability and ecological well-being. A prominent facet of green entrepreneurial endeavors is green innovation, which stands as a proactive strategy to both safeguard the environment and achieve financial gains.

Tien et al. (2020) cited Frederick (2018) explain that moving past conventional commercial and social entrepreneurship, which contribute value to business operations and societal domains respectively, entails enhancing the Earth's value by mitigating the adverse impacts of climate change like global warming and rising sea levels. In a similar vein, eco-friendly entrepreneurship exhibits a heightened awareness of environmental degradation, such as deforestation, and the unsustainable depletion of planetary resources and reserves (Sharma et al., 2022). When concerned about indigenous communities, the majority of them are selling items and providing services identical to them. Basically, their living expenses are borne by selling or showing environment-friendly products or services for the outer community, which represents their engagement in green entrepreneurship activities. Nevertheless, observations and informal conversations with community members have unveiled a concerning trend of rapid decline in indigenous cultural practices and customs, accompanied by the vanishing of the indigenous language and the degradation of the indigenous environmental surroundings. The developmental challenges both within and surrounding the village have made it difficult to evade changes, transformations, or shifts from the traditional indigenous way of life, consequently altering the ecosystem of the Indigenous landscape (Aslam and Gnanapala, 2020)

Moreover, Tien et al. (2020) explain that while doing green-based business green tourism is one component basically practiced by the community of indigenous people. The study done by Aslam and Gnanapala (2020), in the indigenous community within the Vakarei Sri Lanka, explains that the majority of indigenous communities in Sri Lanka engage in tourism activities as income earning opportunity. Moreover, scholars pointed out that while the indigenous community in Vaakarai traditionally relied on hunting, Chena cultivation, and fishing as their primary livelihoods, the current scenario has introduced tourism as a viable alternative source of income. This shift towards tourism has ignited a sense of hopefulness and aspiration within the indigenous community, as they view it as a growing avenue for sustaining their livelihoods. Furthermore, Aslam and Gnanapala (2020) identified that external or non-local entrepreneurs exploit the indigenous resources, profiting from them without adequately compensating the indigenous community. This lack of compensation stems from the community's limited awareness and insufficient capacity and capability in tourism. Despite the initiation of tourism projects and developmental interventions that picture the village to tourists, the indigenous community has not been provided with the appropriate means or opportunities to effectively harness tourism for the renewal of their traditional landscape

3.3   FINDINGS AND DISCUSSION

Both *Dabana* and *Rathugala* indigenous communities' cases show that their livelihood practices are linked to the environment and they make sure that the environment is not harmed by any of their activities. These communities find their salvation, affection, and livelihood by living within the beauty of nature. The village folk stated that generations ago hunting was the main livelihood and there were no other entrepreneurial practices among the communities. They simply use the jungles in the vicinity to hunt and use traditional food-preserving customs by using honey and to make maximum use of crops growing.

However, all means of traditional livelihoods pointed out by the traditional hunter lifestyle were resettled with the post-independence period of Sri Lanka beginning in 1948. Further, the community's nomadic lifestyle and means of subsistence also have been threatened recently by attitude shifts and economic factors. To sustain themselves and their families, many have been compelled to renounce traditional ways of life and look for work often in menial positions that provide neither job security nor dignity. Both chiefs inform that many of the younger clan members are moving to large cities in search of better job chances due to a lack of opportunities, leaving their towns and cultures behind. The chiefs are concerned that the emigration of younger generations and the dearth of employment possibilities in their regions may threaten their culture. When they were being streamlined into society the Veddahs faced numerous threats from the general society, their traditional livelihoods came under enormous threat, and on the other, the continuity of their unique identity was at stake.

3.3.1   Cross Case Analysis

We met our hosts at their cultural centers which aim to showcase their way of life and traditions; (1) The chief of the Rathugala Veddas and his deputy and (2) The Wannila Aththo the leader of the Sri Lankan indigenous community, chief of the Dabana community, and his son. The current study was extended primarily to understand their existing green entrepreneurship practices, strategies, innovative business models, and the challenges they're facing.

3.3.1.1   Green Entrepreneurship Practices and Innovative Business Models

The findings from the interviews and observations revealed that green entrepreneurship practices and innovative business models exist in the Dabana and Rathugala communities as follows.

3.3.1.1.1   *Sales centers*

Both *Dabana* and *Rathugala* communities create, manage, and develop small ventures within their premises for the benefit of their communities (Figures 3.3 and 3.4). These activities encompass both profit-generating activities and those pursued for social reasons to the benefit of the community. However, significant differences between the two communities can be seen in their sales centers.

FIGURE 3.3   Sales center – Rathugala Indigenous Community.

Source: Field Data, 2023e: Field Data, 2023.

FIGURE 3.4 Demonstrating traditional practices.

Source: Field Data, 2023.

Dabana is more oriented toward sales and the variety of goods can be seen in Dabana sales centers than Rathugala community. However, Bee honey is one of the main and common items that can be seen in both sales centers. The fumigation method which is unique and commonly used famous practice by Veddahs to collect bees' honey. The collectors then take the honey they need and leave the excess portion for the bees so they can resume their nest. Interestingly, the collectors had never been stung or attacked during this amazing practice, as were highly confident. Vedda community honey collectors frequently spend many days traveling through the forests to get honey. We truly understand their devotion to the forest, which has provided them with food and care for many generations. The chief claims that over the years, he has worked valiantly to halt destruction and safeguard the forests from unauthorized timber harvesting and agricultural use. Moreover, the Rathugala Veddas make a living by selling beedi leaves (a leaf used to wrap a local cigarette).

The Dabana community is engaged in traditional weaving and handicrafts (Figure 3.3). They sell handmade baskets, jewelry, and crafts made of natural ingredients and wood. However, there is no variation or innovation in the product designs.

3.3.1.1.2 Indigenous Tourism

Over the years, indigenous tourism has advanced significantly, with experienced providers inviting visitors to learn about their traditions and culture. Considering both cases, the Rathugala community is arranging hikes and cave camping experiences for both locals and foreigners with the support of a recently built cultural center. However, the Dabana community is not directly involved with such activities, and they have their own set of traditional activities to attract tourists such as the "Kirikoraha" dance. At present, the Government and some private organizations such as The Culture and Indigenous Communities Programme of Dilmah (PVT) Ltd., Sri Lanka, support indigenous communities in exploring their opportunities in indigenous tourism, traditional craft, and art as a means of empowering the indigenous community by giving them dignity.

3.3.1.1.3 Traditional Activities

As a traditional means of livelihood, the Dabana and Rathugala community presents some of their traditional activities to the visitors such as traditional dance and performances, culinary experience, guiding them for nature walks, and traditional storytelling, etc. Both communities show off their bow-and-arrow skills, traditional dancing, songs, a variety of trap-making techniques, and hunting capabilities. The Deputy of Rathugala Vedda creates a primitive fire from stones to show Stone Age fire-making and earns some amount from the tourists.

Veddah community provides hospitality through a sample of traditional cuisine and beverage experiences. Further beautiful songs and dances accompany many of their rituals and rich traditions. The primary ritual or shantikarma of the indigenous community is the worship of ancestral gods (Ne yakun). People worship their ancestors' deities through traditional performances that include dance and song in order to prevent and heal all illnesses and wounds. Some rituals are carried out yearly, while others have no set timeframe. The three primary rituals are the hakma, kirikora, and hangala dances. All of these ceremonial actions are carried out to ward off diseases, facilitate successful hunting, obtain a plentiful crop of bee honey, and safeguard their chena farming from wild animals.

Even though, it is a part of their cultural inheritance, as a means of livelihood, they are doing demonstrations of these traditional activities (Figure 3.4). Although they earn from traditional shows, in recent times their traditional practices were shaped and reshaped by the changes in Sri Lankan society.

Further, efforts taken to codify the predominant tacit knowledge and knowledge sharing in both cases were limited. Hence, the sustainability of this business model is at risk and there is a requirement to preserve indigenous knowledge for survival and progression before important elements in indigenous knowledge are irretrievably lost. Further, they are nevertheless confident that they would be able to safeguard their environment through moral tourism-related initiatives. Nevertheless, they are optimistic that they would be able to preserve their culture, language, and way of life through moral tourism-related activities, which would also help to advance those ways of life.

3.3.1.1.4   Medicine

Wannila Aththo the chief of Dabana and the Chief of the Sri Lankan indigenous community is famous for doing indigenous medicines. However, the Rathugala community is not directly involved with medicines. They only collect medicines from the forest and sell them to the collectors who are coming from town areas.

Although Vedic remedies were not very sophisticated, the indigenous community did have a variety of methods for treating fractures and other wounds, and all of their drugs were made entirely of herbs. They also believe in the power of spirits to help them heal and recover quickly.

3.3.1.2   Key Challenges and Policy Formulation Insights

3.3.1.2.1   Key Challenges

Indigenous entrepreneurship refers to business initiatives and activities initiated and managed by indigenous people. They are members of diverse cultural, historical, and social communities. They are working hard to develop economic prospects while also protecting and honoring their traditions, land, and heritage. However, because of a history of marginalization, limited access to resources, and cultural differences, they frequently encounter particular challenges when dealing with green entrepreneurial activities.

Accordingly, both Rathugala and Dabana indigenous communities often struggle to access the resources. The major issue that they face is securing the necessary funding to start or grow their green entrepreneurial activities. Poor financial literacy and traditional financing options have made it difficult to find the seed capital to invest in their green innovative practices.

Following that, market accessibility is another key challenge they face. Currently, the indigenous community is following their own green entrepreneurship practices, such as indigenous tourism, selling of environmentally friendly products (value-added products and green products) and services. Thus, they lack access to the mainstream markets since the both Rathugala and Dabana indigenous communities are geographically isolated. Further, the high level of influence of the middlemen in the distribution channels caused a major issue in discouraging these green entrepreneurial activities among the indigenous people.

Further, lack of assistance and collaboration is another growing issue in challenging the green innovative practices of indigenous communities. As an example, the Rathugala indigenous community uses different traditional practices as an income source based on tourism. Thus, the resistance and lack of assistance of the nearby hoteliers is another major issue for the success of their new practices. The poor guidance and the assistance of the authorities also made a significant reason to larger this issue.

Next, the lack of business knowledge and training is identified as a key challenge for indigenous entrepreneurs. A lack of formal education and training in business management and entrepreneurship can be a significant barrier. Indigenous entrepreneurs may not have access to the same resources, networks, communication and educational opportunities as their non-indigenous counterparts.

Moreover, the young generation of the indigenous community resists their traditional practices and tries to mix with the non-indigenous communities. The young generation shows a tendency to seek employment opportunities in urban areas and hesitates to follow traditional entrepreneurial activities in the indigenous community. Since then, this has created a huge challenge of sustaining and growing traditional business activities.

These key challenges hinder the performance of green innovative entrepreneurial practices in indigenous communities. Thus, we suggest the following strategic focuses.

3.3.1.2.2 Strategies to Overcome the Challenges

Green innovation practices in indigenous communities may confront specific hurdles due to cultural, economic, and social considerations. Overcoming these issues necessitates deliberate solutions that take into account the values and aspirations of the community.

Access to Funding and Resources: The grants, funding opportunities, and resources specifically designed for indigenous communities engaged in green innovation. This could come from government programs, philanthropic organizations, or impact investors.

Market Development and Access: Explore the ways to market and promote green products or services to wider audiences. This might involve establishing partnerships with local businesses, tourism initiatives, or online platforms.

Awareness and Education: Raise awareness within the community about the benefits of green innovation practices. Educational campaigns and workshops can help community members understand the value and long-term impact on indigenous communities.

Cultural Alignment: Make certain that green innovation approaches are compatible with indigenous cultural values and customs. Incorporating traditional ecological knowledge and sustainable practices can increase the acceptability and relevance of innovations in the community.

Community Engagement and Ownership: Include the community in the planning and execution of green innovation projects. This promotes a sense of belonging, shared responsibility, and dedication to the efforts.

Capacity Building: Provide training and capacity-building programs to community members in order to provide them with the skills and information required to engage in green innovative practices. This can involve technical abilities, business training, and project management.

Partnerships and Collaboration: Collaborate with green innovation-focused universities, research institutes, non-governmental organizations, and government agencies. Partnerships can provide access to resources such as knowledge, funding, and technology.

Thus, the strategy proposed would provide valuable insights to development practitioners, and policymakers simultaneously to the indigenous community on identifying the potential of green entrepreneurial practices towards sustainable business models and green innovations.

REFERENCES

Asadi, S., OmSalameh Pourhashemi, S., Nilashi, M., Abdullah, R., Samad, S., Yadegaridehkordi, E., Aljojo, N., & Razali, N. S. (2020). Investigating influence of green innovation on sustainability performance: A case on Malaysian Hotel Industry. *Journal of Cleaner Production 258*, 120860.

Aslam, M., & Gnanapala, W. (2020). Merging heritage with rural tourism enterprises in indigenous landscapes; A case study of vaakarai, Sri Lanka.

Attanapola, C. T., & Lund, R. (2013). Contested identities of indigenous people: Indigenization or integration of the Veddas in Sri Lanka. *Singapore Journal of Tropical Geography, 34*(2), 172–187.

Attanapola, C. T., & Lund, R. (2013). Contested identities of indigenous people: Indigenization or integration of the Veddas in Sri Lanka. *Singapore Journal of Tropical Geography, 34*(2), 172–187.

CBSL. (2021). *Economics and Social Statistics of Sri Lanka.* Colombo: Central Bank of Sri Lanka (CBSL).

Dana, L. P. (2015). Indigenous entrepreneurship: An emerging field of research. *International Journal of Business and Globalisation, 14*(2), 158–169.

De Silva, P., & Punchihewa, A. (2011). Socio-anthropological research project on Vedda community in Sri Lanka. *Department of Sociology, University of Colombo: Colombo, Sri Lanka.*

Dharmadasa. (1993). Vedda: the 'Adivasi' of Sri Lanka. *SOBA 4:3, 7–11. Ministry of Environment and Parliamentary Affairs, Sri Lanka.*

Ford, J. D., King, N., Galappaththi, E. K., Pearce, T., McDowell, G., & Harper, S. L. (2020). The resilience of indigenous peoples to environmental change. *One Earth, 2*(6), 532–543.

Frederick, H.H. (2018). The emergence of biosphere entrepreneurship: Are social and business entrepreneurship obsolete? *International Journal of Entrepreneurship and Small Business, 34*(3), 381–419.

Galappaththi, E. K., Ford, J. D., & Bennett, E. M. (2020). Climate change and adaptation to social-ecological change: The case of indigenous people and culture-based fisheries in Sri Lanka. *Climatic Change, 162,* 279–300.

Lansdown, G. (2005).Can you hear me? The right of young children to participate in decisions affecting them. *In Working paper 36.* The Netherlands: Bernard van Leer Foundation.

Lee, K.-H., & Min, B. (2015). Green R&D for eco-innovation and its impact on carbon emissions and firm performance. *Journal of Cleaner Production, 108,* 534–542.

Nikolaoua, E., Ierapetritis, D., & Tsagarakis, K. (2011). An evaluation of the prospects of green entrepreneurship development using a SWOT analysis. *Int. J. Sustain. Dev. World Ecol,18,* 1–16.

Padmasiri, G. R. (2018). Management of indigenous knowledge in Sri Lanka, with special reference to indigenous medicine. *Information development, 34*(5), 475–488.

Senanayake, S. G. J. N. (2006). Indigenous knowledge as a key to sustainable development.

Sharma, N., & Kushwaha, G. (2015). Emerging Green Market as an Opportunity for Green Entrepreneurship and Sustainable Development in India. *Journal of Entrepreneurship and Organizational Management, 4.*

Sharma, R., & Rana, G. (2021). An overview of entrepreneurship development programs in India. *International Journal of Environmental Policy and Decision Making, 3*(1), 16. 10.1504/ijepdm.2021.119511.

Sharma, R., Rana, G., & Agarwal, S. (Eds.). (2022). *Entrepreneurial Innovations, Models, and Implementation Strategies for Industry 4.0* (1st ed.). CRC Press. 10.1201/9781003217084

Tien, N. H., Hiep, P. M., Dai, N. Q., Duc, N. M., & Hong, T. T. K. (2020). Green entrepreneurship understanding in Vietnam. *International Journal of Entrepreneurship, 24*(2), 1–14.

The World Bank Statistics. (2019). Indigenous Peoples.

Tien, N. H., Tien, N. V., Mai, N. P., & Duc, L. D. M. (2023). Green entrepreneurship: a game changer in Vietnam business landscape. *International Journal of Entrepreneurship and Small Business, 48*(4), 408–431.

Veer, C., Kumar, P. & Sharma, R. (2022). Green Entrepreneurship: An Avenue for Innovative and Sustainable Product Development and Performance. In: *Entrepreneurial Innovations, Models, and Implementation Strategies for Industry 4.0.* USA: CRC Press. ISBN 9781032107936.

4 Evaluating Green Human Resource Management Practices in Family-Controlled Hospitality Business

Aditi Sharma, Arun Bhatia, and Mridul

4.1 INTRODUCTION

"Maintaining ecosystem integrity must thus be a primary human goal, which is nevertheless difficult to achieve because little is known about the temporal and spatial scales over which ecosystems should be safeguarded, the limits to replace their functions or the levels of stress they can endure as complex, interacting, and interdependent systems" (p. 283, Gossling, 2002).

The adage "Stay home, Stay safe" is no longer relevant and we are witnessing a global phenomenon of revenge tourism. People are traveling more frequently than ever to experience what they could not during the pandemic. Tourism and hospitality being experience-centered industries need to innovate rapidly to meet the changing expectations of the consumers. Greenwashing is no longer an acceptable solution for satisfying environmentally aware, eco-conscious consumers. Attracting guests who are concerned about sustainability and social responsibility would require the hotels to prioritize sustainable practices and invest in eco-friendly technologies (Mathews, 2023). According to a survey on US hotels, visitors visiting luxury and mid-priced hotels are willing to pay a premium for hotels' green initiatives (Kang, et al., 2012).

Environment Sustainability is gaining massive traction worldwide and businesses can no longer choose to ignore the hot issue. Environmental degradation, depletion of fossil fuels, disturbing and destructing biodiversity and global warming are some of the challenges arising out of the current anthropocentric approach. The tourism industry, one of the largest and fastest-growing economic sectors and a sizable contributor to the global economy, has greatly benefited from the heedless and irresponsible economic activities that have endangered the environment and ecosystem. The exponential increase in tourism-related activities has made the hotel industry one of the biggest global industries (Alreahi et al., 2022). Carbon emissions and the resultant environmental pollution by the hotel and hospitality industry have necessitated the adoption of green and environmentally

DOI: 10.1201/9781003458944-4

friendly practices by the employees working in the industry. Adopting eco-friendly behavior by providing green training to hotel staff can help in promoting environmental commitment and environmental sustainability (Cop et al., 2020). Sustainable tourism practices can aid in the protection and preservation of the natural and cultural environment of tourism destinations (Musa & Nadarajah, 2023). The Hotel Industry in India has embraced a slew of environmental protection measures. For example, Indian Hotels Company Ltd. has implemented a number of environmentally friendly practices, such as reducing carbon emissions, managing waste, and conserving water etc. Furthermore, it has set a benchmark with 78 Earth Check-approved hotels (Sharma & Walia, 2023). In a remarkable achievement; ten ITC Hotels achieved "Net Zero Carbon Status" in December 2022 by achieving the coveted USGBC Zero Carbon certification through sustained efforts. According to the USGBC, ITC Hotels is the world's largest hotel chain with the most LEED Platinum-certified properties. In FY20-21, ITC Grand Chola, a super-premium luxury hotel, became the world's largest LEED Zero Carbon Certified hotel, while ITC Windsor became the world's first (Hospitality World, 2023).

Family businesses are the world's most common type of business organization, and evaluating their environmental performance can help determine how seriously they are incorporating green practices (Miroshnychenko et al., 2022). Family businesses are mainly concentrated in the SME size category around the world and a fraction of them are publicly listed companies (Dekker & Hasso, 2016). Extant literature on family firms reveals that they are laggards in environmental performance as they are more invested in enhancing their financial and personal objectives or socio-emotional endowments (Dal Maso et al., 2020).

Green HRM policies and practices are crucial for gaining competitive advantage, organizational commitment, green supply chain management, environment manage-ment and attainment of sustainable goals of any business (Faisal & Naushad, 2020; Ansari et al., 2021; Rana & Arya 2023). Pro-environment behavior can be fostered among employees through adopting green HRM and can lead to the better environ-mental performance of the firm. Although the tourism and hospitality industries contribute significantly to the economy, their negative impact on the environment remains a source of concern (Patwary et al., 2022). Moreover, there is a dearth of studies examining the adoption of green HRM practices by family-owned businesses in the tourism and hospitality industry (Tuan, 2022) in emerging economies.

The chapter is organized as follows: In the following section, the theoretical framework linking GHRM and the environmental performance of FCBs in the hospitality sector is demystified. Thereafter, what specific Green HRM practices are effective in family-controlled hospitality units and how the commitment of family members affects the implementation of green practices have been examined. The barriers to the adoption of GHRM by FCBs follows in the next section. The discussion and conclusions are presented in the later sections.

It is suggested that embracing green practices such as green work-life balance, green performance management, training and development focused on developing green competencies by family-based firms can be strong pillars for building a sustainable tourist and hospitably unit. Thus, it can be concluded that incorporating green HRM practices into a company's competitive strategy can help it attract and keep talented and ethical employees as well as leverage its green brand image to draw in an increasing number of eco-aware customers.

4.2   OBJECTIVES OF THE STUDY

The aim of this study is to provide insights into the prevalent GHRM practices and their influence in promoting environmental sustainability in family-owned hospitality firms. Furthermore, the challenges in embracing GHRM practices are studied. Finally, the advantages of adopting GHRM are positioned to positively influence the decision-making process of the top echelons in FCBSs.

4.3   METHODOLOGY

The authors reviewed existing literature on Green HRM practices employed by FCB units. Electronic Databases for the literature search were Scopus and Google Scholar. Owing to the fact that there is very little on GHRM practices in the context of family-controlled business in tourism and hospitality, we also reviewed the reference lists given in the studies.

4.4   CONCEPTUAL FRAME WORK

Conceptual framework is proposed in Figure 4.1

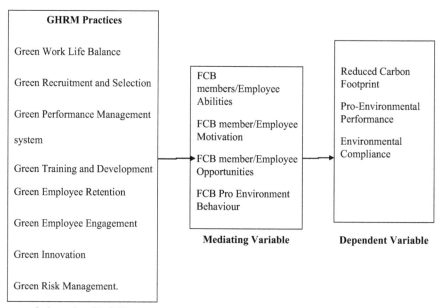

FIGURE 4.1   Conceptual framework.

Source: Author's own.

Despite the fact that most Asian and emerging countries heavily rely on the tourism and hospitality industry, academic research on the effect of their activities on the environment is still underexplored (Nguyen et al., 2022). In recent years consumers, government regulations and civic society have created enough duress on the organizations to practice environment management (Uslu, et al., 2023). Various theories (e.g. stakeholder theory, resource-based theory, ability-motivation-opportunity theory, social cognitive theory) have been used to study GHRM practices. According to ability-motivation-opportunity theory, HRM practices affect the employees' abilities (through training and development), motivation (through work-life balance, performance management, retention) and opportunities (through employee engagement, innovation). Moreover, GHRM practices can positively affect green behavior (PEB) among employees (Ansari et al., 2021). Nonetheless, the effective implementation of GHRM practices depends on the recruitment of employees whose values align with the environmental values and beliefs of the organization (Uslu, et al., 2023). In the conceptual framework, the author proposes the underlying mechanism and processes that link GHRM and its outcomes (reduced carbon footprint, pro-environmental performance, and environmental compliance) in the family-controlled business units in the tourism and hospitality industry.

4.5 THE THEORETICAL LINKAGE BETWEEN GREEN HRM AND ENVIRONMENTAL PERFORMANCE IN THE HOSPITALITY INDUSTRY

Green HRM refers to the use of HRM practices that enable and promote the integration of environment-focused behaviors with the Human resource management function (Shafaei, Nejati, & Yusoff, 2020). The inclusion of environment-orientated goals in a business's general strategy might not be the most effective means to reach sustainability outcomes (Ababneh, 2021). Thus, "Green HRM" offers the use of every employee touch point to act sustainably and also utilize the potential of the human resources in achieving pro-environment outcomes through improved awareness, engagement and commitment towards Green goals (Mandip, 2012). The same, when applied in family-owned hospitality businesses, would demonstrate the firm's consciousness and commitment towards the environment.

Despite growing recognition of the significance of employee's environmentally conscious actions in enhancing the firm's environmental performance, the existing literature on GHRM and its role in the hotel industry remains less explored (Kim et al., 2019). The industry puts considerable strain on the environment. However, such effects can be minimized as Fatoki (2021) suggested that the hospitality industry, through proper environment-management strategies and staff training, can strengthen environmental orientation and therefore generate a Green competitive advantage.

The individual employee's abilities are not replicable and therefore authors, Albrecht et al. (2015) have posited that strategic emphasis on HRM functions can provide a unique competitive advantage to the firm. Additionally, the emphasis on green HR functions would create a Green competitive advantage derived from the employees who have developed their ability and motivation to adopt eco-friendly initiatives at work (Kim et al., 2017).

Furthermore, Darvishmotevali and Altinay (2022) explored the link between GHRM and pro-environmental performance. The authors found support for the GHRM practices leading to increased environmental awareness among hotel employees, which relates positively to proactive pro-environment performance. Singal (2014) asserted that major hospitality businesses are family-controlled and their aim to safeguard the firm's reputation and ensure its longevity, is reflected through their efforts in CSR activities. CSR is closely tied to a firm's strategy of cultivating a sense of environmentally conscious efforts or green orientation (Tanveer et al., 2023). The overall aim of GHRM is a reduction of the firm's ill effects on the environment and an increase in the firm's positive effects on the environment (Arulrajah et al., 2016). However, scholars have not adequately specifically reviewed the unique selling proposition of green practices for family-owned hospitality businesses.

Moving on to the next discussion, the major theories that have linked GHRM and environmental performance are explicated.

4.5.1 STAKEHOLDER THEORY

The stakeholder theory propounds the idea that businesses must create value for all types of stakeholders (Dmytriyev et al., 2021). This theory underscores the role of stakeholders in the adoption of GHRM practices by a firm. For example, today, a customer, a major stakeholder in the hospitality industry, is aware of the effects of the firm's operations on the environment. An environment-aware customer is likely to choose a service provider who is associated with conscious and more environmentally responsible behaviors (Tanova & Bayighomog, 2022). This view was empirically tested by Guerci et al. (2016). The authors found that pressure created by customers and regulatory stakeholders was positively linked to the environmental performance of the firm. This relationship was mediated by GHRM, where the use of green practices by the firm (like Green training and involvement) led to improved environmental performance. Thus, by incorporating environment-conscious actions, the firms take control of their carbon footprints.

4.5.2 RESOURCE-BASED THEORY

The resource-based view suggests that an organization can acquire a sustained competitive advantage through its resources. The majority of studies have discussed the Green competitive advantage acquired through the unique contribution of Human resources developed through the Resource-based view. In the hospitality business, human resources are central to the firm's success. It is, through their creativity, interaction and unique relationship with the consumers, that the business blooms. For example, Farooq et al. (2022) reported the role of GHRM practices in promoting Green creativity in hospitality firms. This view supports that the incorporation of Green practices in HRM function provides a unique, competitive edge to the firms.

4.5.3 THE ABILITY-MOTIVATION-OPPORTUNITY (A-M-O) THEORY

This theory in the GHRM context suggests that for employees to display green behaviors, they must have the required ability, motivation and they must be given the opportunities to act sustainably (Appelbaum et al., 2000; Iftikar et al., 2022; Tanova and Bayighomog, 2022). Abualigah et al. (2022) argued that different GHRM practices like green training and development enhance employee ability, green performance appraisal and awards lead to motivation to act in an eco-friendly manner and green involvement provides opportunities for the employee to work towards sustainable outcomes. Thus, among the hotel employees, the authors found that GHRM practices fostered green engagement at work and it further led to enhanced green creativity among employees. It has been used in studies in the hospitality sector by authors such as Pham et al. (2020) to establish the link between GHRM and the firm's environmental performance.

4.5.4 SOCIAL COGNITIVE THEORY

Drawing on this theory, it is proposed that the employees become more conscious of the impact of their actions on the environment by learning skills to add to the firm's environmental performance (Bandura, 2014; Darvishmoteval and Altinay, 2022). So, the implementation of GHRM practices first makes the employees aware of the impact of their actions on the environment. Through awareness, the employee then recognizes the importance of such practices and engages in Proactive Pro-environmental performance. Abbas et al. (2023) based on SCT, proposed that the use of GHRM practices will lead to a sense of responsibility towards the environment among the employees. This is likely to promote an environment-favorable culture and thus, better social as well as environmental performance. Their findings supported the relationship of GHRM with green culture in the organization. This creation of green culture in the hotel industry then promotes environmental performance. Tuan (2022) integrated the learning theory and the HRM system strength perspective (Ostroff & Bowen, 2016) to investigate how Green HRM practices are linked to the Organizational Citizenship behavior (towards the environment) of the employees. The sample comprised Chinese and Vietnamese hospitality sector employees. The author posited the crucial role of the HR system in ensuring implementation of the needed Green HR practices. These practices can lead to the creation of responsible leadership among managers which is reflected in terms of stakeholder-oriented values. Employees are likely to feel a responsibility towards the environment by observing the environmentally respon-sible behavior of their leader. Through the empirical study, the author found evidence for the role of a leader in transmitting Green HR practices clearly and thus, supporting the employees in internalizing the required value system and obligation towards the environment.

In conclusion, the existing theories provide strong theoretical evidence for the positive outcomes of GHRM practices for all types of corporations including family-controlled enterprises. The next section examines the green practices in family-owned firms.

4.6   GHRM PRACTICES IN FAMILY-CONTROLLED HOSPITALITY BUSINESSES: CHALLENGES AND OPPORTUNITIES

In the tourism and hospitality industries, it is widely acknowledged that small, owner-operated businesses are the dominant model, although there are some multinational organizations that do collaborate within these fields (Getz et al., 2004). Many companies today adopt the "corporate greening model" strategy. This approach prioritizes market-based activities that cater to the increasing demand for eco-friendly products and services. It also takes into account ethical as well as sustainability-oriented values and legal considerations that promote green initiatives (Post & Altman, 1994). The 2008 study by Tzschentke et al. confirms that family-controlled firms in the hospitality industry prioritize environmental measures as part of their values. Moreover, the 2011 study by Tsang identifies work attitudes, moral discipline, status, and relationships as crucial factors for providing better quality services in the hospitality and tourism sector. Family-controlled businesses in the hospitality industry recognize the significance of their brand, which stems from their family values. Shared values among family members, along with personal, sociocultural, and situational factors, are critical to the success of these businesses, given the changing public opinion towards corporate responsibility.

Since travel and tourism are largely human-centric in family-controlled hospitality businesses and are more family-driven, the major focus to undermine success in the field of tourism and hospitality comes from the dynamics of human resource management and its practices. In the hospitality industry, family-run businesses place a strong emphasis on environmentally friendly practices, which they consider a core value.

According to Mwita (2019), the implementation and support of Green HRM practices in an organization can result in significant financial and non-financial benefits. These practices prioritize task performance and the acquisition of knowledge and skills for environmentally friendly behavior. Implementing Green HRM practices in family-owned businesses can significantly reduce carbon footprints through individual actions and initiatives. Such practices would not only promote and enhance green concepts but also evaluate and reward green ideas in the performance systems of family members and the business. Failure to implement these practices can result in severe negative consequences for the environment and the overall success of the business. Therefore, it is imperative that family-owned businesses take immediate action towards implementing green HRM practices.

When considering the operations of family-controlled businesses (FCBs) within the tourism and hospitality sectors and the implementation of eco-friendly business practices, a plethora of ideas and concepts arise. These musings aim to assist in determining whether Small and Medium-Sized Enterprises (SMEs) will adopt these notions. To better understand the concept of GHRM among SME's multitude of subdomain areas need to be evaluated in terms of Green Work-Life Balance (GWLB), Green Recruitment and Selection, Green Performance Management system (GPMS), Green Training and Development, Green Employee retention, Green Employee engagement, Green Innovation, Green Risk Management.

4.6.1   Green Work-Life Balance

Achieving a balance between work and personal life is crucial for overall life satisfaction, as per Robak et al. (2016). In their book "The Family Business in Tourism and Hospitality" (2004), Getz, Carlsen, and Morrison point out that family-controlled businesses in the hospitality industry often struggle with work-life balance as family members tend to work long hours while managing both personal and professional responsibilities. Smaller companies find it harder to implement work-life programs as they require more resources and a stronger focus on organizational culture. It's pertinent to mention that family businesses are becoming increasingly interested in work-life crafting. It emphasizes personal resources such as mental and physical well-being, recognizing that taking care of oneself is crucial for achieving success at work (Dreyer & Busch, 2022). Arora and Wagh's 2017 research, which used Wegmans as an example of a family-owned business, explored how companies can promote work-life balance. The study revealed that offering part-time employees flexible schedules gives them a sense of autonomy and importance.

4.6.2   Green Recruitment and Selection

Recruiting and selecting employees can be difficult for family-controlled hospitality businesses, as documented in the literature. In these types of enterprises, family members or employees act as brand advocates for the family's values and corporate mission. Finding the right person for the job is crucial for the development of the company's vision. However, family-controlled hospitality businesses try to minimize costs by avoiding expenses related to hiring, training, and administration. **Green Performance Management System:** The performance management system focuses on employee actions that promote environmental protection and evaluates the outcomes of green recruitment and selection. Tandon et al. (2023) identified three main activities under the umbrella of "green HRM" based on the works of Renwick et al. (2013) and Pham et al. (2020). These activities include encouraging employees to demonstrate their green skills, motivating employees to use their green skills, and providing opportunities for employees to implement green practices. Through consistent monitoring and assessment, the company recognizes and rewards green initiatives taken by employees and their families, highlighting their dedication to environmental protection.

4.6.3   Green Training and Development

When considering training and development in the 21st century, it's important to weigh the advantages and disadvantages. In order to support environmentally-friendly business practices, the significance of training and development has increased in terms of its impact on job performance. According to Herberich (1998), environmental protection training modules are crucial for the tourism industry due to their systematic and multidisciplinary approach. Additionally, using examples to convey ideas is an effective training method. In their 2020 research article titled "Family Ownership and Environmental Performance: The Mediation

Effect of Human Resource Practices," Dal Maso et al. found that training and workforce development are vital solutions for improving environmental performance. According to Pham et al. (2019) research, it is unequivocally evident that hospitality industry workers who undergo green training are highly inclined to embrace environmentally friendly practices.

4.6.4   Green Employee Retention

The hospitality industry is highly susceptible to employee turnover, indicating the urgent need for organizations in this sector to address this issue (Dogru et al., 2023). To ensure long-term success, companies must provide certification programs that improve both their employees' work and personal lives. Neglecting to do so can lead to high staff turnover rates, which can harm the company's overall prosperity. Likhitkar and Verma (2017) recommend a recognition-based approach that includes initiatives such as family recognition programs and special leave policies to retain skilled workers. Therefore, companies should prioritize the implementation of these plans to secure sustainable success. Managing employee retention in family-run businesses is crucial for maintaining effective management of the company. To ensure the continued success of the business, it's important to retain family members who possess valuable skills and experience. This not only benefits the current generation but also paves the way for future generations of the family to become owners if the business continues to prosper.

4.6.5   Green Employee Engagement

Family businesses have a strong sense of culture and tradition, which can affect the engagement level of employees, as they may perceive a stronger emotional connection with the organization. The dual concern theory also suggests that engagement levels of employees are high when the employees feel that the organization is concerned about their well-being and performance(Rana & Sharma 2019). The study emphasizes the importance of leadership, communication, and job satisfaction in increasing employee engagement and sustainability in family businesses(Ramirez-Lozano et al., 2023). Employee engagement in green initiatives is determined by the willingness of employees to respond to economic, emotional, and social resources and support perceived at work. GHRM practices have a positive association with employee engagement, and effective leadership plays a vital part in strengthening it in the hospitality industry (Ali Ababneh et al., 2021). Furthermore, Ashraf et al.'s study (2015) revealed that workers who are labeled as "green" are more motivated, engaged and enthusiastic about exhibiting green behavior.

4.6.6   Green Innovation

Green innovation is critical for the sustainable performance of hotels since it can reduce the negative environmental impacts of the hospitality industry (Asadi et al., 2020). In a study of accommodation managers operating in Catalonia, Spain, Garay et al. (2019) underscored the relevance of belief systems in the practice of sustainability-oriented innovations. Eco-friendly initiatives are a significant component of innovation for tourist companies and destinations that aim to mitigate harmful environmental impacts and manage stakeholder relationships (Satta, 2019). Although the firms in tourism and hospitality are focusing on green innovation to address environmental issues, build brand awareness generate revenue strategies, there is a pressing need for future research in green innovation (Mert& Mehmet, 2021). Furthermore, a systematic literature review of 61 articles on green and sustainable innovation in tourism too revealed that it is still a fractured research topic requiring further exploration (Satta, 2019).

4.6.7 GREEN RISK MANAGEMENT

Memili and Koç (2023) suggested that small and medium-sized hotels run by families in Turkey could increase resilience and decrease threats to business continuity. The authors emphasized the importance of taking proactive measures to minimize workplace hazards and prioritize the health of employees. In challenging times, it is essential for family members to maintain their business's resilience. Creating an environmentally friendly workplace can help manage the risks associated with family-run hospitality units, particularly in food and beverage areas.

4.7 BARRIERS TO ADOPTION OF PRO-ENVIRONMENTAL BEHAVIOR BY FAMILY-OWNED FIRMS

SMEs, particularly family firms, confront major hurdles in sustaining their businesses and retaining competent staff, particularly in emerging economies (Ramirez-Lozano, J. et al., 2023). The in-depth investigation of the literature on family-controlled businesses in the hospitality and tourism industries highlights two major issues. Apparently, all family members are engaged in the business as employees, so professional human resource management techniques cannot be useful in a family-controlled business (Getz, Carlsen & Morrison, 2004). Second, a greater use of the word "green" on a global scale has led to an understanding of the "green philosophy of business" by most organizations, leading to better use of environmentally friendly techniques and an enhancement of the "green business value chain" (Hasan et al., 2019). However, small family-controlled hospitality units admire the concept but put it least into practice in terms of identifying the key areas to manage and operate green ideology in their business practice. Additionally, there are no external forces that compel small family-run hospitality business units to adopt green practices. According to Kim et al. (2017) adoption of green practices in tourism and hospitality units is difficult since it depends on internal organizational reasons, stakeholder demand, and laws resulting from environmental pressure. The existing literature highlights that exogenous factors like green consumers, green products, and green marketing (Gelderman et al., 2021) are primarily focused on achieving the profit maximization goal resulting from pressure exerted by the external business environment. Thus, very little implementation is found in family-controlled hospitality units like green HR Practices because it is found to be more internally driven by family values.

It is crucial to acknowledge that workplace hazards can significantly impact an employee's job performance. Camilleri and Valeri's (2021) recent study highlights the multiple risks that small family-owned businesses face from external factors, which can endanger their employees' personal, professional, financial, and technological well-being. Detecting poisonous gases and smoke during daily work or job duties can make the workplace less hospitable and raise concerns about the environment. A better operational plan in the green domain, along with an identified environmental management system, strategic environmental proactivity, implementation of green philosophy, and last but not least, better work conditions in small family-controlled hospitality units, can prevent the situation of being at risk and can prevent businesses from going out of business very soon.

4.8 DISCUSSION AND CONCLUSION

The tourism and hospitality sector is dependent on the natural beauty of the area and ironically, it is one of the most polluting industries. The natural ecosystem of popular tourist locations is severely tweaked by the careless and flagrant abuse of natural resources. Embracing green business HRM practices can significantly advance the sustainability efforts of the family-controlled hospitality unit and result in a greening of our planet. Green HRM practices can help these organizations gain a sustainable advantage over their competitors and help them create a distinct brand image. Innovative green solutions are required to address the problems these businesses confront, including their limited capitalization, knowledge of their target markets, promotional activities, and rivalry with larger organized enterprises (Pizam & Upchurch, 2002; Getz et al., 2004). Green HRM practices like green work-life balance, green recruitment and selection, green performance management system, green training and development, and green employee retention (Mwita, 2019) facilitate the congruence of the organization's goals with sustainability objectives. However, in family-dominated hospitality firms, achieving a green work-life balance—that entails achieving harmony between one's personal and professional life—remains a challenge. This is due to the incongruence between the goals of family firms and green objectives.

Implementing green hiring and selection practices can strengthen an organization's commitment to the environment and stimulate a green workplace culture (Ashraf et al., 2015). The evaluation and acknowledgment of employees' eco-friendly behaviors and actions is a key component of the implementation of a green performance management system. Family-controlled hospitality enterprises can cultivate a sense of pride and motivation among staff members by recognizing and rewarding green initiatives, leading to a more committed and engaged workforce (Pham et al., 2020). In conjunction with training and development activities, green performance management can further enable staff to acquire the skills and knowledge required to contribute to the organization's environmental initiatives (Pham et al., 2019). Green practice-focused training and development programmes play a critical role in providing staff with the knowledge they need to advance sustainability inside the company (Herberich, 1998). Nevertheless, fostering green competencies training continues to be a challenge in family-controlled enterprises

(Dal Maso et al., 2020). However, it is pertinent to reiterate that research on GHRM indicates that funding green training may have a positive effect on staff involvement, resulting in a rise in voluntary green behavior (Pham et al., 2019).

Family-controlled enterprises are not an exception when it comes to the difficulty in retaining employees (Dogru et al., 2023). Green HRM practices can increase employee retention rates and promote a more dedicated and environmentally conscientious staff by recognizing and rewarding green innovation and offering possibilities for growth and development (Likhitkar & Verma, 2017; Fazal-e-Hasan et al., 2023). Notwithstanding the challenges, implementing Green HRM practices is beneficial for the environment and the businesses as a whole. Family-run firms may stand out from the competition, draw in clients who care about the environment, and support a more sustainable tourist and hospitality sector. Some of the shining examples of hospitality units located in India that are family-managed include Spice Village (Kerala), Dune Eco Village and Spa (Puducherry), Neemrana Hotel, Banjara Camps and Retreats (Himachal Pradesh). Organic farming, rainwater harvesting, using solar power to meet their energy needs and minimal use of single-use plastics are some of the examples of sustainability initiatives being undertaken by them and have become their USPs.

In the 21st century, adopting Green HRM with emphasis on environmental responsibility is not only an ethically appropriate corporate strategy but also a legal requirement. In conclusion, Family-run hospitality business units can make a significant contribution to protecting the environment for future generations and reducing the industry's total ecological effect by committing to sustainable practices.

REFERENCES

Ababneh, O. M. A. (2021). How do green HRM practices affect employees' green behaviors? The role of employee engagement and personality attributes. *Journal of Environmental Planning and Management, 64*(7), 1204–1226. 10.1080/09640568.2020.1814708.

Abbas, Z., Smaliukienė, R., Zámečník, R., Kalsoom, G., & Cera, E. (2023). How does green HRM influence environmental and social sustainability in hotels? *Problems and Perspectives in Management, 21*(1), 253–263. 10.21511/ppm.21(1).2023.22.

Abualigah, A., Koburtay, T., Bourini, I., Badar, K., & Gerged, A. M. (2022). Towards sustainable development in the hospitality sector: Does green human resource management stimulate green creativity? A moderated mediation model. *Business Strategy and the Environment*, 10.1002/bse.3296.

Albrecht, S. L., Bakker, A. B., Gruman, J. A., Macey, W. H., & Saks, A. M. (2015). Employee engagement, human resource management practices and competitive advantage: An integrated approach. *Journal of Organizational Effectiveness: People and Performance, 2*(1), 7–35. 10.1108/JOEPP-08-2014-0042.

Alreahi, M., Bujdosó, Z., Kabil, M., Akaak, A., Benkó, K. F., Setioningtyas, W. P., & Dávid, L. D. (2022). Green human resources management in the hotel industry: A systematic review. *Sustainability, 15*(1), 99. 10.3390/su15010099.

Ansari, N. Y., Farrukh, M., & Raza, A. (2021). Green human resource management and employees pro-environmental behaviours: Examining the underlying mechanism. *Corporate Social Responsibility and Environmental Management, 28*(1), 229–238. 10.1002/csr.2044.

Appelbaum, E., Bailey, T., Berg, P., et al. (2000). *Manufacturing advantage: Why high performance systems pay off.* Ithaca, NY: Cornell University Press.

Arora, C., & Wagh, R. (2017). Importance of work-life balance. *International Journal of New Technology and Research, 3*(6), 23–25.

Arulrajah, A. A., Opatha, H. H. D. N. P., & Nawaratne, N. N. J. (2016). Green human resource management practices: A review. *Sri Lankan Journal of Human Resource Management, 5*(1), 1–16.

Asadi, S., Pourhashemi, S. O., Nilashi, M., Abdullah, R., Samad, S., Yadegaridehkordi, E., ... & Razali, N. S. (2020). Investigating influence of green innovation on sustainability performance: A case on Malaysian hotel industry. *Journal of Cleaner Production, 258,* 120860.

Ashraf, F., Ashraf, I., & Anam, W. (2015). Green HR for businesses. *International Journal of Academic Research in Business and Social Sciences, 5*(8), 149–156. 10.6007/IJARBSS/v5-i8/1771.

Bandura, A. (2014). Social cognitive theory of moral thought and action. In *Handbook of moral behavior and development* (pp. 69–128). Psychology press. 10.4324/9781315807294.

Banjara Experiences: Offbeat Himalayan stays & adventures. Retrieved August 4, 2023 from https://banjaraexperiences.com/.

Camilleri, M. A., & Valeri, M. (2021). Thriving family businesses in tourism and hospitality: A systematic review and a synthesis of the relevant literature. *Journal of Family Business Management, 12*(3), 555–576. 10.1108/JFBM-10-2021-0133.

Cop, S., Alola, U. V., & Alola, A. A. (2020). Perceived behavioral control as a mediator of hotels' green training, environmental commitment, and organizational citizenship behavior: A sustainable environmental practice. *Business Strategy and the Environment, 29*(8), 3495–3508. 10.1002/bse.2592.

Dal Maso, L., Basco, R., Bassetti, T., & Lattanzi, N. (2020). Family ownership and environmental performance: The mediation effect of human resource practices. *Business Strategy and the Environment, 29*(3), 1548–1562. 10.1002/bse.2452.

Darvishmotevali, M., & Altinay, L. (2022). Green HRM, environmental awareness and green behaviors: The moderating role of servant leadership. *Tourism Management, 88,* 104401. 10.1016/j.tourman.2021.104401.

Dekker, J., & Hasso, T. (2016). Environmental performance focus in private family firms: The role of social embeddedness. *Journal of Business Ethics, 136,* 293–309. 10.1007/s10551-014-2516-x.

Dmytriyev, S. D., Freeman, R. E., & Hörisch, J. (2021). The relationship between stakeholder theory and corporate social responsibility: Differences, similarities, and implications for social issues in management. *Journal of Management Studies, 58*(6), 1441–1470. 10.1111/joms.12684.

Dogru, T., McGinley, S., Sharma, A., Isık, C., & Hanks, L. (2023). Employee turnover dynamics in the hospitality industry vs. the overall economy. *Tourism Management, 99,* 104783. 10.1016/j.tourman.2023.104783.

Dreyer, R., & Busch, C. (2022). At the heart of family businesses: How copreneurs craft work-life balance. *Journal of Family Business Management, 12*(4), 816–832. 10.1108/JFBM-12-2020-0113.

Dune Eco Village & Spa Hotel; Dune eco village and spa (2016, April 10). Retrieved from August 2, 2023, from https://dunewellnessgroup.com/dune-eco-village-spa/.

Faisal, S., & Naushad, M. (2020). An overview of green HRM practices among SMEs in Saudi Arabia. *Entrepreneurship and Sustainability Issues, 8*(2), 1228–1244. 10.9770/jesi.2020.8.2(73).

Fatoki, O. (2021). Environmental orientation and green competitive advantage of hospitality firms in South Africa Mediating effect of green innovation. *Journal of Open Innovation: Technology, Market, and Complexity, 7,* 223. 10.3390/joitmc7040223.

Farooq, R., Zhang, Z., Talwar, S., & Dhir, A. (2022). Do green human resource management and self-efficacy facilitate green creativity? A study of luxury hotels and resorts. *Journal of Sustainable Tourism*, *30*(4), 824–845. 10.1080/09669582.2021.1891239.

Fazal-e-Hasan, S. M., Ahmadi, H., Sekhon, H., Mortimer, G., Sadiq, M., Kharouf, H., & Abid, M. (2023). The role of green innovation and hope in employee retention. *Business Strategy and the Environment*, *32*(1), 220–239. 10.1002/bse.3126.

Garay, L., Font, X., & Corrons, A. (2019). Sustainability-oriented innovation in tourism: An analysis based on the decomposed theory of planned behavior. *Journal of Travel Research*, *58*(4), 622–636. 10.1177/0047287518771215.

Gelderman, C. J., Schijns, J., Lambrechts, W., & Vijgen, S. (2021). Green marketing as an environmental practice: The impact on green satisfaction and green loyalty in a business-to-business context. *Business Strategy and the Environment*, *30*(4), 2061–2076. 10.1002/bse.2732.

Getz, D., Carlsen, J., & Morrison, A. (2004). *The family business in tourism and hospitality*. CABI.

Gössling, S. (2002). Global environmental consequences of tourism. *Global Environmental Change*, *12*(4), 283–302. 10.1016/S0959-3780(02)00044-4.

Guerci, M., Longoni, A., & Luzzini, D. (2016). Translating stakeholder pressures into environmental performance–The mediating role of green HRM practices. *The International Journal of Human Resource Management*, *27*(2), 262–289. 10.1080/09585192.2015.1065431.

Hasan, M. M., Nekmahmud, M., Yajuan, L., & Patwary, M. A. (2019). Green business value chain: A systematic review. *Sustainable Production and Consumption*, *20*, 326–339. 10.1016/j.spc.2019.08.003.

Herberich, A. (1998). Environmental Training for Tourism Professionals. *European Environmental Management Association*, France.

Hospitality World, E. T. (2023, March 10). *ITC contributes to the PM's Panchamitra Strategy with its green building initiatives*. ET Hospitality. World. https://hospitality.economictimes.indiatimes.com/news/hotels/itc-contributes-to-the-pms-panchamitra-strategy-with-its-green-building-initiatives/98545245

Iftikar, T., Hussain, S., Malik, M. I., Hyder, S., Kaleem, M., & Saqib, A. (2022). Green human resource management and pro-environmental behaviour nexus with the lens of AMO theory. *Cogent Business & Management*, *9*(1), 2124603. 10.1080/23311975.2022.2124603

Kang, K. H., Stein, L., Heo, C. Y., & Lee, S. (2012). Consumers' willingness to pay for green initiatives of the hotel industry. *International Journal of Hospitality Management*, *31*(2), 564–572. 10.1016/j.ijhm.2011.08.001.

Kim, S. H., Lee, K., & Fairhurst, A. (2017). The review of "green" research in hospitality, 2000-2014: Current trends and future research directions. *International Journal of Contemporary Hospitality Management*, *29*(1), 226–247. 10.1108/IJCHM-11-2014-0562.

Kim, Y. J., Kim, W. G., Choi, H. M., & Phetvaroon, K. (2019). The effect of green human resource management on hotel employees' eco-friendly behavior and environmental performance. *International Journal of Hospitality Management*, *76*, 83–93. 10.1016/j.ijhm.2018.04.007.

Likhitkar, P., & Verma, P. (2017). Impact of green HRM practices on organization sustainability and employee retention. *International Journal for Innovative Research in Multidisciplinary Field*, *3*(5), 152–157.

Mandip, G. (2012). Green HRM: People management commitment to environmental sustainability. *Research Journal of Recent Sciences*, ISSN, 2277, 2502.

Mathews, B. (2023, January 9). *Hospitality Industry in 2023: Responding to new trends*. Hospitality Net. Retrieved July 24, 2023, from https://www.hospitalitynet.org/opinion/4114325.html.

Memili, E., & Koç, B. (2023). The antecedents of family firms' resilience to crisis in hospitality and tourism. *International Journal of Hospitality Management, 113*, 103526. 10.1016/j.ijhm.2023.103526.

Miroshnychenko, I., De Massis, A., Barontini, R., & Testa, F. (2022). Family firms and environmental performance: A meta-analytic review. *Family Business Review*. 10.11 77/08944865211064409.

Musa, F., & Nadarajah, R. (2023). Valuing visitor's willingness to pay for green tourism conservation: A case study of Bukit Larut Forest Recreation Area, Perak, Malaysia. *Sustainable Environment, 9*. 10.1080/27658511.2023.2188767.

Mwita, K. (2019). Conceptual review of green human resource management practices. *East African Journal of Social and Applied Sciences, 1*(2), 13–20.

Neemrana hotels. (n.d.). Neemranahotels.com. Retrieved August 4, 2023, from https://www.neemranahotels.com/.

Nguyen, N., Nguyen, H. V., D'Souza, C., & Strong, C. (Eds.). (2022). *Environmental sustainability in emerging markets: Consumer, organisation and policy perspectives*. Springer. ISBN: 978-981-19-2407-1.

Ostroff, C., & Bowen, D. E. (2016). Reflections on the 2014 decade award: is there strength in the construct of HR system strength? *Academy of Management Review, 41*(2), 196–214. 10.5465/amr.2015.0323.

Patwary, A. K., MohdYusof, M. F., Bah Simpong, D., AbGhaffar, S. F., & Rahman, M. K. (2022). Examining proactive pro-environmental behaviour through green inclusive leadership and green human resource management: An empirical investigation among Malaysian hotel employees. *Journal of Hospitality and Tourism Insights*. 10.1108/JHTI-06-2022-0213.

Pham, N. T., Hoang, H. T., & Phan, Q. P. T. (2020). Green human resource management: A comprehensive review and future research agenda. *International Journal of Manpower, 41*(7), 845–878. 10.1108/IJM-07-2019-0350.

Pham, N. T., Thanh, T. V., Tučková, Z., & Thuy, V. T. N. (2020). The role of green human resource management in driving hotel's environmental performance: Interaction and mediation analysis. *International Journal of Hospitality Management, 88*, 102392. 10.1016/j.ijhm.2019.102392.

Pham, N. T., Tučková, Z., & Jabbour, C. J. C. (2019). Greening the hospitality industry: How do green human resource management practices influence organizational citizenship behavior in hotels? A mixed-methods study. *Tourism Management, 72*, 386–399. 10.1016/j.tourman.2018.12.008

Pizam, A., & Upchurch, R. S. (2002). The training needs of small rural tourism operators in frontier regions. *Tourism in Frontier Areas*. Lexington Books.

Post, J. E., & Altma, B. W. (1994).Managing the environmental change process: Barriers and opportunities. *Journal of Organizational Change Management, 7*(4), 64–81. 10.1108/09534819410061388.

Ramirez-Lozano, J., Peñaflor-Guerra, R., & Sanagustín-Fons, V. (2023). Leadership, communication, and job satisfaction for employee engagement and sustainability of family businesses in Latin America. *Administrative Sciences, 13*(6), 137. 10.3390/admsci13060137.

Rana, G., & Sharma, R. (2019). Emerging human resource management practices in Industry 4.0. *Strategic HR Review, 18*(4), 176–181.

Rana, G., & Arya, V. (2023), Green human resource management and environmental performance: mediating role of green innovation – A study from an emerging country, *Foresight*. 10.1108/FS-04-2021-0094.

Renwick, D. W., Redman, T., & Maguire, S. (2013). Green human resource management: A review and research agenda. *International Journal of Management Reviews, 15*(1), 1–14.10.1111/j.1468-2370.2011.00328.x.

Robak, E., Słocińska, A., & Depta, A. (2016). Work-life balance factors in the small and medium-sized enterprises. *PeriodicaPolytechnica Social and Management Sciences*, *24*(2), 88–95.

Satta, G., Spinelli, R., & Parola, F. (2019). Is tourism going green? A literature review on green innovation for sustainable tourism. *Tourism Analysis*, *24*(3), 265–280. 10.3727/1 08354219X15511864843803.

Shafaei, A., Nejati, M., & MohdYusoff, Y. (2020). Green human resource management: A two-study investigation of antecedents and outcomes. *International Journal of Manpower*, *41*(7), 1041–1060. 10.1108/IJM-08-2019-0406.

Sharma, K., & Walia, M.K. (2023, February 21). *Green Initiatives in hotel industry towards achieving 'Saptarishi'–green growth goal of budget 2023*. ET Hospitality World. https://hospitality.economictimes.indiatimes.com/news/speaking-heads/green-initiatives-in-hotel-industry-towards-achieving-saptarishigreen-growth-goal-of-budget-2023/98110622.

Singal, M. (2014). Corporate social responsibility in the hospitality and tourism industry: Do family control and financial condition matter? *International Journal of Hospitality Management*, *36*, 81–89. 10.1016/j.ijhm.2013.08.002. world

Spice Village – Resort in Thekkady. (n.d.). CGHEarth. Retrieved August 4, 2023, from https://www.cghearth.com/spice-village.

Tandon, A., Dhir, A., Madan, P., Srivastava, S., & Nicolau, J. L. (2023). Green and non-green outcomes of green human resource management (GHRM) in the tourism context. *Tourism Management*, *98*, 104765. 10.1016/j.tourman.2023.104765.

Tanova, C., & Bayighomog, S. W. (2022). Green human resource management in service industries: The construct, antecedents, consequences, and outlook. *The Service Industries Journal*, *42*(5–6), 412–452. 10.1080/02642069.2022.2045279.

Tanveer, M. I., Yusliza, M. Y., & Fawehinmi, O. (2023). Green HRM and hospitality industry: Challenges and barriers in adopting environmentally friendly practices. *Journal of Hospitality and Tourism Insights*. 10.1108/JHTI-08-2022-0389.

Tsang, N. K. (2011). Dimensions of Chinese culture values in relation to service provision in hospitality and tourism industry. *International Journal of Hospitality Management*, *30*(3), 670–679. 10.1016/j.ijhm.2010.12.002.

Tuan, L. T. (2022). Promoting employee green behavior in the Chinese and Vietnamese hospitality contexts: The roles of green human resource management practices and responsible leadership. *International Journal of Hospitality Management*, *105*, 103253. 10.1016/j.ijhm.2022.103253.

Uslu, F., Keles, A., Aytekin, A., Yayla, O., Keles, H., Ergun, G. S., & Tarinc, A. (2023). Effect of green human resource management on green psychological climate and environmental green behavior of hotel employees: The moderator roles of environmental sensitivity and altruism. *Sustainability*, *15*(7), 6017. 10.3390/su15076017.

5 Impact of eWOM on Green Cosmetics Purchasing Intentions
An Emerging Market Perspective

Ramzan Sama and Ravindra Sharma

5.1 INTRODUCTION

The rapidly increasing consumer consciousness towards the environment has changed consumer behaviour and preferences (Alwitt & Pitts, 1996; Rana & Arya 2023). It has increased the demand for green products (Kong et al., 2014). Green purchase behaviour is posited to be important in a sustainable environment (Joshi & Rahman, 2016). In fact, we must consider green products to minimise the adverse environmental effects (Chandan Veer et al. 2022; Quoquab & Mohamad, 2017). Thus, it is important to understand green consumer behaviour (Yadav & Pathak, 2017).

Despite consciousness of global environmental problems, not all consumers exhibit green purchasing habits (Quoquab & Mohamad, 2019). The extant literature indicates that positive attitudes or intentions toward green products necessarily leads to green purchasing (Bruschi et al., 2015; Tanner & Kast, 2003). According to D'Souza & Taghian (2005). In the case of cosmetics products, consumers' demand for eco-friendly products is increasing. However, in developing countries, not all consumers prefer green products (Jaini et al., 2020), which calls for further inquiry to understand green purchasing behaviour (Liobikienė et al., J. 2017; Quoquab et al., 2019a). The present study fills this critical gap.

Due to the rise in the demand from consumers, the sales of cosmetics have increased significantly. Furthermore, Asia Pacific region has the biggest market share in the cosmetics industry with three billion consumers (Zion et al., 2018). Due to environmental problems, manufacturers make green cosmetics from natural ingredients that are free from side effects (Acme-Hardesty, 2019). Green Cosmetics "are cosmetic products made from natural ingredients produced from renewable raw materials". The importance of green cosmetics is on the rise

DOI: 10.1201/9781003458944-5

because of its multiple benefits to the environment, health and society (Women With Mind, 2019). Thus, this study focuses on the Asian market, India and the green cosmetics category.

The increased use of social media has drastically influenced consumer behaviour (Wang, 2017). It gives a platform to share consumers' opinions in the form of Electronic Word-of-Mouth (eWOM) communication. Consequently, affecting consumers brand perceptions (Hennig-Thurau et al., 2004). According to Duan et al. (2008) eWOM "is to the exchange of product-related information, opinions, and experiences through digital platforms, such as social media, online forums, and review websites". The extant research has studied the impact of eWOM on purchase decisions (Shang et al., 2017; Vahdati & Nejad, 2016). However, its impact on green purchasing intentions needs to be explored. It is important as most consumers refer to others' opinions on social media. Hence, this study posits eWOM as an antecedent to green purchasing intentions. Accordingly, this explores the mediating role of attitude and green concerns between eWOM and green purchasing intentions in the context of the cosmetics industry, which is a relatively unexplored area.

The basic presumption in this research is that consumers' attitudes and green concerns play a vital role in influencing their green purchasing intentions for cosmetics products. The impact of eWOM and attitude has been investigated in the literature (Al-Halbusi & Tehseen, 2018) in different contexts. Attitude is defined by Fishbein and Ajzen (1975) as a "consumer's overall evaluation of a product or brand, encompassing both cognitive and affective components" (Fishbein & Ajzen, 1975). However, more research needs to be done in the green cosmetics context. Whereas, green concerns are an individual's commitment towards green products. Based on this discussion, the stimulus-organism-response (SOR) theory, the theory of reasoned action (TRA) and the theory of planned behaviour (TPB), this study proposed the relationship between eWOM and attitude towards green cosmetics and green concerns further led to green purchasing intentions.

EWOM influences consumers' attitudes and behaviour (Zhang et al., 2017). In line with this research and SOR theory, this study explores the interactions between eWOM, attitude, green concerns and green purchasing intentions for green cosmetics. By understanding the mediating effect of attitude towards green cosmetics and green concerns between eWOM and green purchasing intentions, this study unveils consumer insights into green purchasing intentions. The findings guide researchers and practicing managers in the cosmetics industry, unveiling deeper insights into incorporating eWOM in marketing communications for green products.

In the next section, the theoretical underpinning of the research context (SOR theory, TRA and TPB), extant literature and conceptual model and proposition are discussed, followed by the conclusion, and theoretical and managerial discussion. Finally, the limitations and future scope of the study are discussed.

Proposed Conceptual Model: The proposed conceptual model has been shown in Figure 5.1

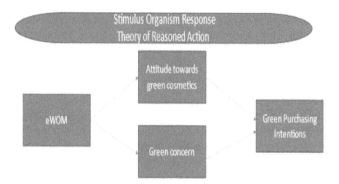

FIGURE 5.1   Proposed model.

Sources: Author's own.

5.2   LITERATURE REVIEW AND PROPOSITION DEVELOPMENT

5.2.1   THEORETICAL BACKGROUND

Grounded on the SOR theory (Mehrabian & Rusell, 1974), this study is considered the basis for the proposed conceptual framework. This theory indicates that stimulus affects individual organisms, leading to either positive or negative responses (Quoquab et al., 2019b). The extant research has indicated eWOM as the stimulus, brand attitude and cognitive response as the organism, and purchase intentions as the response (Yan et al., 2018; Emir et al., 2016).

Extant research (Kalafatis et al., 1999; Lada et al., 2009) has employed the TRA and TPB (Ajzen, 1985, 1991) to investigate consumers' purchase intention across product categories including environmentally friendly products. TRA and TPB explains the links between attitude, behaviour and intentions. The TRA-guided attitude leads to intentions, whereas TPB posits perceived behavioural control that affects intentions. Thus, this study considers TRA and TPB to explore the relationship between eWOM, attitude and green purchasing intentions in the context of green cosmetics.

5.2.2   ELECTRONIC WORD OF MOUTH (EWOM) AND ATTITUDE TOWARDS GREEN COSMETICS

According to Ajzen (1991), an attitude refers to "the extent to which an individual makes a favourable or unfavourable evaluation of behaviour towards an object". A source's credibility plays a vital role in determining eWOM authenticity (Cheung & Thadani, 2012). Since eWOM are from consumers it is perceived as authentic and credible, leading to favourable consumer attitudes (Hennig-Thurau et al., 2004). People base their decisions for specific behaviours on the cost-benefit analysis of performing that behaviour (Quoquab et al., 2017). Additionally, eWOM about green products sensitise consumers to considering their purchases' impact on the environment (Zhu & Zhang, 2010). The extant research has indicated a positive effect of eWOM on consumers' attitudes and behaviour (Li & Hitt, 2008; Reichelt et al., 2014). These studies were in various contexts, such as retail and fashion products in online and offline settings. In this line, this study assumes that online information can influence consumers' attitudes towards green cosmetics positively, resulting in green purchasing intentions. This assumption is based on SOR theory and TRA. Thus, it proposed the following:

Proposition 1: There is a relationship between eWOM and attitude towards green cosmetics

5.2.3 εWOM and Green Concerns

Past research has established that consumers are becoming more conscious about the environment, leading to changes in consumer behaviour in general and purchase behaviour in particular (Papadopoulos et al., 2010). Compared to advertisers, consumers put more trust in peer consumers (Sen & Lerman, 2007). It has been established that WOM positively affects attitude and behavioural intentions (Chatterjee, 2001). Internet is one of the important communication channels; thus, in line with this discussion, this study posits the relationship between eWOM and consumers' environmental concerns. Guided by this discussion and in line with SOR theory and TRA and TPB, the following proposition has been proposed:

Proposition 2: There is a relationship between eWOM and green concerns

5.2.4 εWOM and Green Purchasing Intentions

The opinions and recommendations of others affect consumers' preferences (Zhao & Xie, 2011). Thus, eWOM is one of the important marketing strategies employed by marketers. Chen et. al. (2014) posit that eWOM reduces consumer uncertainty and thus helps in decision-making. Keller and Fay (2012) established that positive WOM leads to high credibility. Information overload across media platforms about green products confuses consumers. Hence WOM positively affects green purchasing intentions (Chen et al., 2011). In line with past researchers and grounded on SOR theory, we posit that eWOM affects green purchasing intentions. Thus, we proposed

Proposition 3: There is a relationship between eWOM and green purchasing intentions

5.2.5 Attitude Towards Green Cosmetics and Green Purchasing Intentions

The theory of reasoned action (TRA) and the TPB (Ajzen, 1985, 1991) established the impact of attitude on behaviour. These theories posit that consumers either approach or avoid the object based on positive/negative feelings towards the object. Extant research has also established the positive impact of attitude on behaviour (Chekima et al.,2016) and Chetioui et al., 2020). In an exciting study, Kalafatis et al. (1999) propose the positive effect of attitude on green purchasing intentions. This study posited that TRA, grounded on SOR theories, is in line with this.

Proposition 4: Relationship between Attitude towards green cosmetics and Green purchasing intentions

5.2.6   GREEN CONCERNS AND GREEN PURCHASING INTENTIONS

Extant research has established the relationship between green concerns and green purchasing intentions (Ottman et al., 2006; Newton et al., 2015). These studies indicated that a degree of environmental awareness leads to green purchasing behaviour. However, limited studies have tested this relationship in green cosmetics perspectives. Biswas and Roy (2015) posit the positive effect of green concerns on green purchasing intentions. Further, due to increased awareness about the environmental effects of non-green cosmetics, demand for green cosmetics has increased significantly. Thus, consumers prefer green cosmetics over non-green (Chan & Lau, 2018). In an interesting study, Kim and Choi (2019) established a positive relationship between consumers' green concerns and their intention to buy green cosmetics. Therefore, researchers argue that green concerns lead to green purchasing intentions. Accordingly, the following proposition has been proposed:

Proposition 5: There is a relationship between Green concerns and Green purchasing intentions

5.2.7   ATTITUDE AS A MEDIATOR BETWEEN eWOM AND GREEN PURCHASING INTENTIONS

Extant research has established the mediating role of attitude in consumer research (Sama & Trivedi, 2019; Mohammad et al., 2021). For example, in the context of wireless finance, Amin et al. (2017) established the mediating role of attitude. In addition, SOR theory supports this proposition, for instance, eWOM (stimulus) that affects consumers' emotions positively/negatively (organism), further leading to green purchasing intentions (response). Since environmental concerns drive green cosmetics purchasing, eWOM can affect attitudes further, leading to green purchasing intentions. On the basis of this discussion, the following proposition has been proposed:

Proposition 6: Attitude towards green cosmetics mediates the relationship between eWOM and green purchasing intentions

5.2.8 MEDIATING ROLE OF GREEN CONCERNS

It is found that consumers with high levels of green concerns prefer more eco-friendly products (Biswas & Roy, 2015). In an interesting study by Johnstone and Tan (2015), Chinese consumers with a high degree of green concerns reduce their irresponsible consumption. Limited studies have focused on exploring green concerns' mediating role between online information and green purchasing intentions. Based on this guidance and SOR theory, we assume in this study that eWOM stimulus led to organism green concern, further leading to response green purchasing intentions. Thus, research proposes that green concerns act as a mediator between eWOM and green purchasing intentions.

Proposition 7: Green concerns mediate the relationship between eWOM and green purchasing intentions

5.3 CONCLUSION

Across the globe, the rising concerns about the environment have made consumers conscious of their consumption of products, leading to more emphasis on green products. Still, all consumers do not prefer green purchasing. Past research indicated a positive relationship between attitude and green purchasing intentions, but in the case of cosmetics products, developing countries consumers do not prefer green purchasing (Jaini et al., 2020), which requires further investigation (Liobikienė et al., J. 2017; Quoquab et al., 2018). This study focuses on green cosmetics and addresses this gap. Based on a concise literature review and grounded on SOR theory, TRA, TPB, this study proposes several propositions related to eWOM and green purchasing intentions.

Due to the advancement of information technology, consumer information exchange on social media platforms has increased considerably. This eWOM has positively impacted the consumers' attitude towards green cosmetics and also increased the green concerns. The extant research has established the positive relationship between eWOM and attitude towards green cosmetics and green concerns. This research proposed the relationship between eWOM and attitude towards green cosmetics and green concerns. Further, this study proposed the mediating role of attitude and green concerns between eWOM and green purchasing intentions in the context of green cosmetics. This proposition is based on SOR theory, which guided eWOM as the stimulus, and attitude and green concerns as the organism leading to the response of green purchasing intentions. Additionally, in line with TRA and TPB, which suggest that attitude leads to behaviour – in this context attitude towards green cosmetics and green concerns – the cognitive and emotional aspects lead to green purchasing intentions.

5.4 CONTRIBUTION TO THEORY AND PRACTICE

This study contributes to the theory from two perspectives. First, this study proposed the mediating role of attitude and green concerns between eWOM and green purchasing intentions. This is an interesting contribution, as limited research focused on green cosmetics products in emerging markets like India. Second, this study proposes the relationship between eWOM and attitude towards green cosmetics and green concerns, further leading to green purchasing intentions.

From a practical perspective, managers can take a clue from this and incorporate eWOM in their marketing strategies as it has direct and indirect relationships with green purchasing intentions. Hence, green cosmetics marketers need to focus on eWOM as one of the important marketing communications strategies so that it impacts attitudes in a positive fashion. As eWOM takes place on social media platforms, marketers should use these platforms to spread awareness, benefits and concerns about green cosmetics products.

5.5   LIMITATION AND FUTURE SCOPE

This study has several limitations. This study is from the perspective of developing countries. Hence, the same proposition cannot be tested in developed countries. This study has suggested propositions based on a concise literature review. Future studies can test this proposition and may establish the positive or negative relationship between eWOM and attitude and green concerns in the case of green cosmetics. Further research can empirically test the mediating effect of attitude and green concerns between eWOM and green purchasing intentions. Moreover, this study focuses on green cosmetics, future studies may propose the eWOM relationship with the same constructs with different product categories such as fashion and electronic gadgets, to name a few.

REFERENCES

Acme-Hardesty (2019). Green cosmetics: the push for sustainable beauty. Available at: www.acme-hardesty.com/green-cosmetics-sustainable-beauty/ (accessed 21 January 2019).

Al-Halbusi, H. & Tehseen, S. (2018). The effect of electronic word-of-mouth (eWOM) on brand image and purchase intention: a conceptual paper. *Socio-Economic Challenges*, 2(3), 83–94.

Alwitt, L. F. & Pitts, R. E. (1996). Predicting purchase intentions for an environmentally sensitive product. *Journal of Consumer Psychology*, 5(1), 49–64.

Amin, H., Abdul Rahman, A. R., Abdul Razak, D. & Rizal, H. (2017). Consumer attitude and preference in the Islamic mortgage sector: a study of Malaysian consumers. *Management Research Review*, 40(1), 95–115

Biswas, A. & Roy, M. (2015). Green products: an exploratory study on the consumer behaviour in emerging economies of the East. *Journal of Cleaner Production*, 87, 463–468.

Bruschi, V., Shershneva, K., Dolgopolova, I., Canavari, M. & Teuber, R. (2015). Consumer perception of organic food in emerging markets: evidence from Saint Petersburg, Russia. *Agribusiness*, 31(3), 414–432.

Chan, R. Y. & Lau, L. B. (2018). A study of the attitude towards the use of green cosmetic products: a Hong Kong perspective. *Asia Pacific Journal of Marketing and Logistics*, 30(1), 229–241.

Chatterjee, P. (2001). Online reviews: Do consumers use them? *Advances in Consumer Research*, 28, 129–133.

Chen, Y., Wang, Q. & Xie, J. (2011). Online social interactions: a natural experiment on word of mouth versus observational learning. *Journal of Marketing Research,* 48(2), 238–254.

Chen, Y. S., Lin, C. L. & Chang, C. H. (2014). The influence of greenwash on green word-of-mouth (green W.O.M.): the mediation effects of green perceived quality and green satisfaction. *Quality & Quantity,* 48(5), 2411–2425.

Cheung, C. M. K. & Thadani, D. R. (2012). The impact of electronic word-of-mouth communication: A literature analysis and integrative model. *Decision Support Systems*, 54, 461–470. 10.1016/j.dss.2012.06.008.

D'Souza, C., & Taghian, M. (2005). Green advertising effects on attitude and choice of advertising themes. *Asia Pacific Journal of Marketing and Logistics*, 17, 51–66. 10.11 08/13555850510672386.

Duan, W., Gu, B., & Whinston, A. (2008). The dynamics of online word-of-mouth and product sales—An empirical investigation of the movie industry. *Journal of Retailing*, 84, 233–242. 10.1016/j.jretai.2008.04.005.

Emir, A., Halim, H., Hedre, A., Abdullah, D., Azmi, A. & Kamal, S. B. M. (2016). Factors influencing online hotel booking intention: a conceptual framework from stimulus-organism-response perspective. *International Academic Research Journal of Business and Technology*, 2(2), 129–134.

Fishbein, M. & Ajzen, I. (1975). *Belief, attitude, intention, and behavior: an introduction to theory and research.* Boston: Addison-Wesley.

Hennig-Thurau, T., Gwinner, K. P., Walsh, G., & Gremler, D. D. (2004). Electronic word-of-mouth via consumer-opinion platforms: What motivates consumers to articulate themselves on the Internet? *Journal of Interactive Marketing*, 18, 38–52. 10.1002/dir.10073.

Jaini, A., Quoquab, F., Mohammad, J. & Hussin, N. (2020). "I buy green products, do you … ?" The moderating effect of eWOM on green purchase behavior in Malaysian cosmetics industry. *International Journal of Pharmaceutical and Healthcare Marketing*, 14(1), 89–112.

Johnstone, M. L. & Tan, L. P. (2015). Exploring the gap between consumers' green rhetoric and purchasing behaviour. *Journal of Business Ethics*, 132, 311–328.

Joshi, Y. & Rahman, Z. (2016). Predictors of young consumer's green purchase behavior. *Management of Environmental Quality: An International Journal*, 27 (4), 452–472.

Kalafatis, S. P., Pollard, M., East, R. & Tsogas, M. H. (1999). Green marketing and Ajzen's theory of planned behaviour: a cross-market examination. *Journal of Consumer Marketing*, 16(5), 441–460.

Keller, E. & Fay, B. (2012). Word-of-Mouth advocacy. *Journal of Advertising Research,* 52 (4).

Kim, Y. & Choi, S. M. (2019). Effects of green marketing on consumers' purchase intentions in the organic cosmetics context. *Journal of Retailing and Consumer Services*, 49, 139–144.

Kong, W., Harun, A., Sulong, R. S. & Lily, J. (2014). The influence of consumers perception of green products on green purchase intention. *International Journal of Asian Social Science*, 4(8), 924–939.

Lada, S., Tanakinjal, G. H. & Amin, H. (2009). Predicting intention to choose halal products using theory of reasoned action. *International Journal of Islamic and Middle Eastern Finance and Management*, 2(1), 66–76.

Li, X., & Hitt, L. M. (2008). Self-selection and information role of online product reviews. *Information Systems Research*, 19, 456–474. 10.1287/isre.1070.0154.

Liobikienė, G. & Bernatonienė, J. (2017). Why determinants of green purchase cannot be treated equally? The case of green cosmetics: literature review. *Journal of Cleaner Production*, 162, 109–120.

Mehrabian, A., & Russell, J. A. (1974). The basic emotional impact of environments. *Perceptual and Motor Skills*, 38, 283–301. 10.2466/pms.1974.38.1.283.

Mohammad, J., Quoquab, F. & Mohamed Sadom, N. Z. (2021). Mindful consumption of second-hand clothing: the role of eWOM, attitude and consumer engagement. *Journal of Fashion Marketing and Management: An International Journal*, 25(3), 482–510.

Newton, J. D., Tsarenko, Y., Ferraro, C. & Sands, S. (2015). Environmental concern and environmental purchase intentions: The mediating role of learning strategy. *Journal of Business Research*, 68(9), 1974–1981.

Ottman, J. A., Stafford, E. R. & Hartman, C. L. (2006). Avoiding green marketing myopia. *Environment: Science and Policy for Sustainable Development*, 48(5), 22–36.

Papadopoulos, I., Karagouni, G., Trigkas, M., & Platogianni, E. (2010). Green marketing. *EuroMed Journal of Business*, 5, 166–190. 10.1108/14502191011065491.

Quoquab, F. & Mohamad, J. (2017). Managing sustainable consumption: is it a problem or panacea? In Filho, W. L., Pociovalisteanu, D. M. and Al-Amin, A. Q. (Eds.), *Sustainable economic development: green economy and green growth, world sustainability series*, Springer: New York, NY, Chapter 7, pp. 115–125.

Quoquab, F., Mohammad, F. & Sukari, N. N. (2019a). A multiple-item scale for measuring sustainable consumption behavior. *Asia Pacific Journal of Marketing and Logistics*, 31(4), 791–816.

Quoquab, F., Mohamed Sadom, N. Z. & Mohammad, J. (2019b). Driving customer loyalty in the Malaysian fast food industry: the role of halal logo, trust and perceived reputation. *Journal of Islamic Marketing*, 11(6), 1367–1387.

Rana, G. & Arya, V. (2023). Green human resource management and environmental performance: mediating role of green innovation – a study from an emerging country. *Foresight*. 10.1108/FS-04-2021-0094.

Sama, R. & Trivedi, J. P. (2019). Factors affecting consumers' loyalty towards halal cosmetics: an emerging market perspective. *International Journal of Business and Emerging Markets*, 11(3), 254–273.

Sen, S., & Lerman, D. (2007). Why are you telling me this? An examination into negative consumer reviews on the Web. *Journal of Interactive Marketing*, 21, 76–94. 10.1002/dir.20090.

Shang, S. S. C., Wu, Y. L. & Sie, Y. J. (2017). Generating consumer resonance for purchase intention on social network sites. *Computers in Human Behavior*, 69, 18–28

Tanner, C. & Kast, S. W. (2003). Promoting sustainable consumption: determinants of green purchases by Swiss consumers. *Psychology and Marketing*, 20(10), 883–902.

Vahdati, H. & Nejad, S. H. M. (2016). Brand personality toward customer purchase intention: the intermediate role of electronic word-of-mouth and brand equity. *Asian Academy of Management Journal*, 21(2), 1–26.

Veer, C., Kumar, P. & Sharma, R. (2022). Green Entrepreneurship: an avenue for innovative and sustainable product development and performance. In *Entrepreneurial innovations, models, and implementation strategies for Industry 4.0*. CRC Press: USA. ISBN 9781032107936.

Wang, T. (2017). Social identity dimensions and consumer behavior in social media. *Asia Pacific Management Review*, 22(1), 45–51.

Women With Mind (2019). Sustainability beauty – green cosmetics. Available at: www.womanwithmind.com/sustainability-green-cosmetics/ (accessed 19 October 2019).

Yadav, R. & Pathak, G. S. (2017). Determinants of consumers' green purchase behavior in a developing nation: applying and extending the theory of planned behavior. *Ecological Economics*, 134, 114–122.

Yan, X., Shah, A., Zhai, L., Khan, S. & Shah, A. (2018). Impact of mobile electronic word of mouth (eWOM) on consumers purchase intentions in the fast-casual restaurant industry in Indonesia. *Proceedings of the 51st Hawaii International Conference on System Sciences, Honolulu, Hawaii*, 8, 3801–3810, doi: 10.24251/HICSS.2018.479.

Zhang, T., Abound Omran, B. & Cobanoglu, C. (2017). Generation Y's positive and negative eWOM: use of social media and mobile technology. *International Journal of Contemporary Hospitality Management*, 29(2), 732–761.

Zhao, M. & Xie, J. (2011). Effects of social and temporal distance on consumers' responses to peer recommendations. *Journal of Marketing Research*, 48 (3), 486–496.

Zhu, F., & Zhang, X. (M.). (2010). Impact of online consumer reviews on sales: The moderating role of product and consumer characteristics. *Journal of Marketing*, 74, 133–148. 10.1509/jmkg.74.2.133.

Zion Market Research (2018). Global cosmetic products market will reach USD 863 billion by 2024. Available at: https://globenewswire.com/news-release/2018/06/22/1528369/0/en/Global-Cosmetic-Products-Market-Will-Reach-USD-863-Billion-by-2024-Zion-Market-Research.html (accessed 25 January 2019).

6 Uncovering the Shifting Landscape

A Comprehensive Analysis of Emerging Trends and Best Practices Adopted in Green Supply Chain Management

V. Harish and Ravindra Sharma

6.1 INTRODUCTION

In the contemporary era, characterized by an unparalleled magnitude of environmental apprehensions, organizations and entities are progressively acknowledging the imperative to adopt sustainable practises. The concept of "Green Supply Chain Management" (GSCM) has gained momentous traction in recent years as a strategic as well as tactical approach to effectively tackle and manage environmental concerns as well as regulations and impeccably adopt sustainability practices and principles into the entire supply chain (Sarkis, Kouhizadeh & Zhu, 2021). This article aims to understand the precise definition, significance, and best practices of GSCM practiced by firms within the framework of modern-day business operations.

The notion of GSCM has reached significant prominence in the past few decades as organizations across the globe aim to address crucial environmental issues and address various stakeholders, such as governments', NGO's, customers', etc., demands for a sustainable way of delivering products and services. In view of the escalating concerns around environmental damage, pollution, changes in weather patterns due to the effect of climate change and natural resource depletion, it is becoming evident that traditional and conventional supply chain practises are no longer deemed adequate and accepted by various stakeholders (Wassie, 2020). In contemporary times, there is an increasingly growing attention and emphasis among organizations spanning diverse sectors to adopt ecologically sustainable practises and principles into their supply chains. This transformative trend of focussing on thinking and acting with the environment in mind is commonly referred to as GSCM.

The incorporation of sustainability into their existing business practises has become critical leading to the advent of GSCM as a strategic approach for

DOI: 10.1201/9781003458944-6

organizations. In view of the increasing concerns with respect to environmental deterioration and the imperative for sustainable progress, researchers and professionals alike have acknowledged the paramount importance of embracing environmental conscious approaches in the supply chain network (Sharma et al. 2022).

This chapter aims to present a clear understanding of the term and evolution of GSCM. The chapter progresses next with a detailed analysis of the earlier literature studies done on the topic discussing the advancements in the topic and various strategies adopted by organizations to augment sustainability in their supply chain operations. The chapter also aims to shed light on understanding the best practices adopted by various organizations and various strategies adopted by organizations across businesses and countries to practice a "green supply chain". The chapter also discusses the recent topics, technologies and various approaches to implementing a "green supply chain" and also aims to present a framework for organizations to implement a "green supply chain".

6.1.1 OBJECTIVE OF THE CHAPTER

- To give a detailed understanding of the evolution of and need for the term GSCM
- To elaborate on the various advancements and innovations in the area of "sustainable and green supply chain"
- To shed light on understanding the best practices adopted by various organizations and various strategies adopted by organizations across businesses and countries to adopt "green supply chain"
- To provide a framework for organizations to implement "green supply chain"

6.2 METHODOLOGY

The foundational guidelines of PRISMA "Preferred Reporting Items for Systematic Reviews and Meta-Analysis" were adopted for the collection of articles and analysis of the same. The database and search engines used were Scopus, ProQuest, Science Direct and Google Scholar. The timeline of the articles available in the databases was from 1992 to 2023, which is 31 years. Very minimum number of articles that started in the year 1996 and was almost flat till the year 2006. Since 2006 there has been moderate growth in the number of articles and post 2012 from roughly 60 articles per year it has exponentially increased to approximately 240 articles per year (Figure 6.1). As shown in Figure 6.3, a majority of the articles collected were articles/Scholarly journals with the remaining being dissertations, Books, book chapters, conference proceedings and other documents.

The search on the mentioned databases yielded a total of 2463 articles with specific keywords used such as Green supply chain management and sustainable supply chain at the first level, followed by definition, evolution, need and best practices. The consolidated articles were checked for relevancy and recency and only the titles that were recent and relevant were taken for consideration, which resulted in 587 articles. These 587 articles were selected for further literature review and an abstract review revealed that only 451 articles were considered as articles for review for the study on

Documents by year

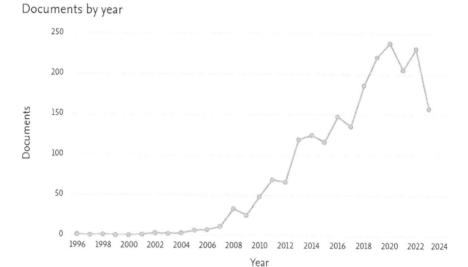

FIGURE 6.1   Year-wise journals/articles reviewed.

GSCM. The inclusion criteria were articles and documents with keywords related to "Green supply chain Management", such as "Sustainable Supply chain", environmental supply chain, etc., at the first stage and for the next stage, certain specific keywords such as "Green manufacturing", "Green Sourcing", Green Logistics", "Green Packaging", "Definition of GSCM", Evolution of GSCM", " "Best practices in GSCM" etc. were used for literature review purposes (Figure 6.2).

Documents by country or territory

Compare the document counts for up to 10 countries/territories.

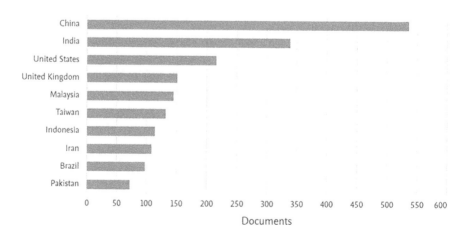

FIGURE 6.2   Documents reviewed by top 10 countries.

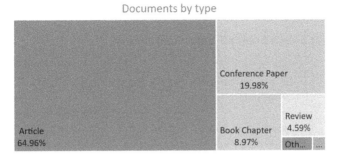

FIGURE 6.3   Documents reviewed by type.

6.3   LITERATURE REVIEW

The following section of the chapter deals with the literature review on the definition, evolution meaning and principles of GSCM.

6.3.1   GREEN SUPPLY CHAIN MANAGEMENT CONCEPT

Green Supply Chain Management (GSCM) denotes the incorporation of environmental and sustainable practices and principles by organizations, from designing the product/service, procurement of raw materials to the distribution of the final product to the end user and proper disposal of the product (AlBrakat, Al-Hawary & Muflih, 2023). GSCM emphasizes the lessening of environmental damage caused as a result of the firm's activities, including the optimal use of resources, reducing carbon emission, proper disposal of waste, etc. (Sugandini et al. 2020). GSCM rather than taking a reactive approach takes a proactive approach in conforming with environmental policies and regulations and aims to address the sustainability obstacles across the supply chain (Park, Kim & Lee, 2022).

6.3.1.1   Defining Green Supply Chain Management

GSCM can be defined as the incorporation of environmentally friendly practices into the entire supply chain process, encompassing the identifying environmentally friendly channel partners, purchasing of raw materials, green production and manufacturing, distribution, and managing the product at the end of its life (Tseng et al. 2019). It involves reducing carbon footprint, minimizing waste generation, sustainable way of disposing of waste, optimizing energy efficiency, and adopting sustainable sourcing practices (Khan et al. 2022). GSCM denotes the use of environmentally friendly practices into the entire supply chain, right from designing the product, procuring raw materials from ideal supply partners to a sustainable production system, optimized distribution, and efficient and safe disposal of the end of the life products. GSCM focuses on balancing environmental considerations with the efficient and effective performance of the supply chain (Leng et al. 2020).

Organizations worldwide have looked into their current supply chain practices to overcome various environmental challenges such as government regulations,

TABLE 6.1

Dimensions of Green Supply Chain Management

Supply Chain	Remarks	Reference Literature
Designing the Product	Ensuring that the firm's components are designed in a "Green way"	Kuiti et al. 2019, Reche et al. 2020, Einizadeh & kasraei, 2021, Uemura Reche et al. 2022
Manufacturing	Ensure that the entire manufacturing process adopts "Green approach"	Aityassine et al. 2021, Awan et al. 2022, García Alcaraz et al. 2022, Sheng et al. 2023
Distribution	Distributing the goods in an optimized way thereby reducing carbon emissions and in a sustainable way	Rajabion et al. 2019, Sant, 2022, Stekelorum et al. 2021, Samekto & Kristiyanti, 2022
Disposal	Having a safe and environmentally friendly and safe disposal mechanism for the products post their life cycle.	Aćimović, Mijušković, & Rajić, 2020, Plaza-Úbeda et al. 2020, Akter, & Saif Abu 2022.

Source: Author's Own

carbon emissions, climate change and sustainable requirements from the customer (Lee, 2015). GSCM has been defined as "a proactive approach to improve the existing processes and products in order to comply with the environmental regulations" (Gelmez, 2020). GSCM has also been defined as "The set of supply chain practices and policies practiced, actions considered and partnerships constituted in order to address the concerns raising from environmental management across designing, manufacturing, distribution and disposal of the firms products and services" (Zsidisin & Sifered, 2001). Srivastava (2007) in their study have defined GSCM as "encompassing the thought of environment into all the supply chain and operational practices such as designing the products, Purchasing, identifying vendors, manufacturing process, distribution of products and services to customers and safe disposal of the products post their product life cycle". Zhu and Sarkis (2007) have defined GSCM as "all the environmental initiative taken across the product life cycle starting from designing, manufacturing, distribution and disposal of the products". The various dimensions of GSCM have been indicated in Table 6.1 as the Design, Manufacture, distribution and disposal stage.

6.3.1.2 Evolution of GSCM

The evolution of the phrase GSCM can be referred to the early '80s when the concern for environmental protection gained significant traction globally. The environmental protection initiatives primarily focused on pollution prevention in land, water and air within individual organizations (Kale, 2021). However, as the awareness and knowledge of environmental dimensions grew, the focus gradually shifted towards across the entire length of the supply chain, recognizing the need for a collaborative and dedicated effort across the supply chain partners is essentially needed to attain marginal environmental improvements. This shift in mindset led to

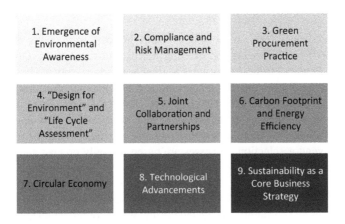

FIGURE 6.4 Evolution of GSCM.

the gaining importance of the concept of GSCM looked at as a holistic, wholesome and integrated approach across the supply chain to achieve sustainability (Menon & Ravi, 2022).

The concept and understanding of the term GSCM have rapidly evolved over the decades, reflecting the mounting awareness of adopting an environmentally sustainable approach in all business functions. This transition in the concept of GSCM can be traced through various vital stages, each stage as shown in Figure 6.4, being marked by evolving challenges, technological advancements, and a greater realization of the need for practicing sustainable practices.

6.3.1.2.1 Emergence of Environmental Awareness

The late '70s and early '80s, saw a gaining momentum in the environmental protection movement, leading to focus on issues such as pollution, deforestation, and depleting natural resources. During this stage, organizations and entities started realizing the negative impact on the nature and environment and started to take initial steps to address the damage to the environment (Dheeraj & Vishal, 1992). This awareness gained traction across the globe and marked the initial steps for environmentally sustainable practices.

6.3.1.2.2 Compliance and Risk Management

The next decade specifically the early '90s saw a transition from voluntary sustainable and environmental initiatives from individual organizations to regulatory compliance being referred by Government acts and policies. Government bodies and agencies began implementing stringent environmental policies and regulations, forcing companies to adopt and comply with pre-defined standards and norms related to waste management, emissions control, and pollution control (Shekari et al. 2011). As a result of the policy measures and tighter control business entities started integrating environmental attentions into their current supply chain practices to avoid reputational and legal risks.

6.3.1.2.3   Green Procurement Practice

The late 90s and early 2000s witnessed a shift in the focus to greening procurement practices apart from following the existing norms and regulations. Organizations realized that along with their existing sustainable practices, other vital aspects of their sustainable efforts were green sourcing and efficient supplier management in order to remain sustainable across their supply chains (Mitra & Datta, 2014). This involved doing supplier assessments to ensure compliance with environmental regulations, encouraging them to adopt sustainable practices along the supply chain and ensuring vendor selection processes are based not only on cost and quality but also on sustainable practices.

6.3.1.2.4   "Design for Environment" and "Life Cycle Assessment"

As organizations and corporations realized the importance of operating in a sustainable way and further reducing their environmental footprint, concepts such as "Design for Environment" (DfE) and "Life cycle Assessment" (LCA) gained popularity. Design for Environment (DfE) focuses on thinking about environmental and sustainable facets in the designing stage itself. It involves thinking about designing the components in such a way that the environmental impact is minimal from the usage of resources from manufacturing to disposal of the component or product (Tseng et al. 2019). LCA is a systematic process to assess and evaluate the environmental damage that a product makes throughout its life from initiation to disposal.

6.3.1.2.5   Joint Collaboration and Partnerships

The turn of the century saw organizations focussing more on joint collaboration and partnerships with their supply chain partners leading to advancements in GSCM. Businesses realized that achieving sustainability could not be achieved in silos but had to be achieved across suppliers, customers, industry associations, and NGOs and started collaborating with them. Collaborative initiatives facilitated the sharing of best practices, Joint cost reduction, and establishing of certifications and industry-wide standards (Yang, 2018).

6.3.1.2.6   Carbon Footprint and Energy Efficiency

With Governments, companies and NGOs' increasing focus on climate change, the emphasis shifted to carbon footprint reduction and energy efficiency in a more scientific and measurable way. Organizations began to evaluate, measure and analyse their carbon footprints across the organization as well as their supply chain partners (Xu et al. 2023). Strategies such as vehicle fleet optimization, intermodal transportation, and route optimization gained importance to minimize carbon emissions in transportation.

6.3.1.2.7   Circular Economy

In the recent past, there has been an increasing awareness and deliberations on the concept of "circular economy" within the operational and supply chain context. The notion of "circular economy" aims to eliminate wasting resources by designing and developing products for durability, reliability, repairability, and recyclability. Organizations as a part of working towards a circular economy have adopted various strategies such as reverse logistics, waste reduction programs, and product take-back initiatives to reduce the environmental damage due to their firms' operations (Liu, Zhu & Feng, 2019).

6.3.1.2.8 Technological Advancements

Rapid progress in technology has had a substantial role in the growth of GSCM. Innovations and technologies such as AI, data analytics, Internet of Things (IoT), and Additive manufacturing have enabled faster, more improved and better decision-making. Technologies such as 4 G/5 G networks, blockchain, etc., (Ravindra et al. 2021) have also augmented supply chain transparency and traceability among supply chain partners, which are critical for sustainability along the supply chain (Sutawijaya & Nawangsari, 2020).

6.3.1.2.9 Sustainability as a Core Business Strategy

Today, the scope and awareness of GSCM have grown exponentially to such a point where sustainability has become an integral part of the core business strategy of organizations. Many companies have realized that sustainability is not just another aspect but has to be a fundamental aspect and should be thought of with a long-term perspective (Despoudi, 2020). Sustainability is now seen as a driver of innovation, providing competitive advantages and tools for improving operational efficiency and organizations have it at the central focus of their business strategy.

6.3.2 MEANING OF THE TERM "GREEN SUPPLY CHAIN MANAGEMENT"

GSCM is a phrase that has gained substantial attention in recent years which denotes to the inclusion of eco-friendly aspects into the operations supply chain management practises (Liu, Chen & Liu, 2020). At its essence, GSCM aims for the synchronization of economic expansion along with the conservation of the environment and natural resources. Researchers have acknowledged that conventional supply chain practises have not considered the impact on the environment, resulting in the faster eroding of natural resources, increased pollution, and thereby degradation of ecosystems as a whole. The focus of GSCM is to provide an optimized and sustainable supply chain system. This can be realized by adopting various initiatives such as reducing carbon footprint, optimizing usage of energy, minimizing waste, and ensuring environmentally responsible suppliers across the chain (Yun et al. 2023).

The term "green supply chain" pertains to the acclimatization of ecologically conscious and sustainable practises and sharing of the best practices in terms of sustainability within the realm of supply chain management. The comprehensive scope of this term encompasses the entirety of a product's lifecycle, commencing with the designing of the products, extraction of raw materials, adopting a green manufacturing, sustainable way of distribution to the end users, and ultimately culminating in the disposal phase at the end stage of the life cycle of the product (Mathu, 2019). The primary objective is to reduce the ecological impact resulting from supply chain activities by optimizing the usage of resources and reducing waste (Krishnan et al. 2020).

The practice of "going green" by individuals, business enterprises and governments has been driven by numerous factors, including but not limited to excessive pollution, high energy consumption, greenhouse gas emissions, an alarming rate of deforestation, increased pollution, and high amount of waste generation (Venhoeven, Bolderdijk & Steg, 2020). The concept of the green supply chain aims to effectively tackle these concerns and foster the progress of sustainable development along with economic growth.

In the domain of production and manufacturing, the concept of a green supply chain revolves around the effective and efficient utilization of sustainable raw materials and the adoption of environmentally conscious production methods (Sheng et al. 2023). The all-inclusive approach encompasses right from the designing stage, the procurement of raw materials from supply partners who strictly adhere to sustainable practises, the optimal and efficient utilization of renewable energy, the reduction or elimination of hazardous raw material substances in the manufacturing process and providing a sustainable and environmentally friendly working place to employees (Karabacak & Saygili, 2022). In addition to their primary focus, manufacturers may also lay emphasis on the development of products that are recyclable or reusable thereby effectively mitigating waste and augmenting the lifespan of their products.

In the domain of transportation and logistics, green supply chains play a critical on the reduction of carbon emissions, route optimization and aiming at transporting materials and goods efficiently (Laguir, Stekelorum & El Baz, 2021). This can be achieved through the adoption of diverse technologies and practices, including the adoption of electric vehicles or choosing alternative fuels, optimizing route planning to minimize traveling distances, incorporating load consolidation strategies. The utilization of cutting-edge and recent advancements in technologies, such as data analytics, driverless vehicles, telematics and real-time tracking, holds immense potential in enhancing transportation efficiency and optimizing fuel usage (Saridogan, 2012).

In the domain of distribution and warehousing, green supply chains involve the adoption of environmentally conscious practices and methodologies aimed at optimizing energy conservation and proper energy utilization (Soyege, Makinde & Akinlabi, 2023). Organizations may adopt various strategies such as energy-efficient lighting systems, recycling programmes, practicing sustainable packaging systems, and the adoption of sustainable practises in inventory management. Furthermore, it is imperative for companies to probe into the area of utilizing renewable energy sources, such as using solar panels, in order to have their warehouses and distribution centers more sustainable (Nyachomba & Achuora, 2022).

Another crucial dimension of GSCM pertains to the integration between the suppliers on one side and customers on the other side. This can be achieved by the establishment of collaborative partnerships with the focus being on sustainability with both suppliers and customers. Organizations will have to establish joint collaborative relationships with their suppliers in order to achieve sustainable sourcing, ensure that their suppliers adhere to ethical labor conditions, and comply with environmental regulations. The integrated effort across supply chain partners may encompass the sharing of best practices, knowledge transfer, conducting audits, and the enablement of supplier development initiatives (Mathu, 2019). The active involvement of customers/ end users in the adoption of sustainable practises, such as the promotion of recycling initiatives or the implementation of take-back programmes for products during their end-of-life stages, constitutes a key aspect within the framework of a green supply chain.

The implementation of a green supply chain has the potential to yield a multitude of positive outcomes for organizations. Some of the positive outcomes for corporates are bolster corporate sustainability and social responsibility, thereby enhancing the company's reputation and brand image. Furthermore, the implementation of environmentally friendly practises often results in significant financial benefits through energy conservation, waste minimization, and streamlined and efficient operational processes and procedures (Liu, Chen & Liu, 2020).

6.3.3   THE NEED FOR GREEN SUPPLY CHAIN MANAGEMENT

In today's rapidly evolving world, the need for GSCM has become more demanding than in earlier years. As much as businesses strive to be profitable and show continuous growth, they must also remember the environmental challenges that pose a severe threat to the well-being of our planet. GSCM offers a tactical as well as a strategic approach to address these concerns by integrating sustainability into every aspect of the supply chain, from designing, sourcing, manufacturing, delivery, and disposal. The following points emphasize the need for organizations to adopt GSCM practices as shown in Figure 6.5.

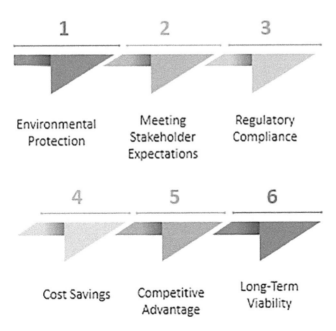

FIGURE 6.5   Need for green supply chain management.

6.3.3.1   Environmental Protection

One of the primary drivers for organizations to adopt green supply chain management is the pressing need to protect and save the environment. With wide concerns hanging over topics such as climate change, deforestation, and pollution, there is an increasing need for organizations to ensure to reduce their environmental footprint (Kale, 2021). GSCM enables firms to minimize negative environmental impact, conserve natural resources, and contribute to a better planet.

6.3.3.2   Meeting Stakeholder Expectations

Stakeholders such as customers, suppliers, end users, investors, governments, etc., are increasingly demanding the need to adopt more sustainable and environmentally friendly practices from organizations (Chkanikova & Sroufe, 2021). Adopting GSCM allows firms to meet these expectations for the stakeholders, thereby enhancing their reputation, attracting environmentally conscious customers and partners, and finally ensuring that the stakeholder trusts them.

6.3.3.3   Regulatory Compliance

Governments across the globe are enforcing stringent environmental policies, regulations and standards. Non-compliance with these policies and regulations can result in penalties, legal consequences, and damage to a company's reputation (Wei, 2023). Adopting GSCM helps organizations adhere to these requirements, ensuring legal compliance and minimizing potential risks.

6.3.3.4   Cost Savings

With firms aiming to be cost-competitive to survive, implementing green practices across the supply chain can lead to significant cost savings and help them remain profitable. GSCM practices such as the usage of energy-efficient technologies, initiatives to eliminate waste, and streamlined manufacturing and transportation systems can lower operational costs considerably over time (Gunawardena, 2021). GSCM allows businesses to optimize resource utilization, minimize wastage and improve overall cost efficiency.

6.3.3.5   Competitive Advantage

Embracing sustainability by adopting GSCM can provide organizations with a competitive edge over competitors. Businesses that proactively adopt environmentally sustainable practices have a unique selling proposition and thereby differentiating themselves from their competitors and attracting environmentally conscious customers, creating new business opportunities and mitigating the risk of non-compliance (Al-khawaldah et al. 2022).

6.3.3.6   Long-Term Viability

Sustainable practices not only help the firm to comply with the regulations but also assist firms to remain competitive in the long term (Sugandini et al. 2020). Embracing GSCM ensures that organizations are better prepared for uncertain future challenges, including government policy changes, changing consumer requirements and most importantly managing resource scarcity.

6.4   KEY PRINCIPLES OF GREEN SUPPLY CHAIN MANAGEMENT

The building blocks of GSCM can be classified into the following seven guiding principles as indicated in Figure 6.6 are as follows:

1. Environmental Regulatory Adherence: Organizations must ensure regulatory adherence by complying with environmental regulations and standards specified by Governments throughout their supply chain activities (Kale, 2021). This principle confirms that the organization's operational activities meet legal statutory requirements and minimize negative impacts on the environment.
2. Sustainable Sourcing: Practicing sustainable sourcing involves in identifying and selecting suppliers who adhere to environmentally sustainable practices in their supply chain activities. This principle focuses on procuring raw materials and services from suppliers who practice sustainability in their entire supply chain network from procurement to manufacturing (Adhikari, Biswas & Avittathur, 2019).
3. Waste Reduction and Recycling: One key aspect of GSCM is to implement strategies to ensure that waste generation is reduced and recycling is welcomed to the maximum extent possible (Alrubah et al. 2020). This principle involves adopting measures to ensure that waste is minimized throughout the supply chain by adopting Lean management principles, optimizing packaging design, applying recycling programs, and opting for circular economy approaches.
4. Energy Efficiency: Energy efficiency is a key principle of GSCM to ensure the conservation of natural resources. Organizations should aim to optimize energy consumption and sustainability and increase the use of

FIGURE 6.6   Key principles of GSCM.

Source: Author's Own.

renewable energy sources across every stage of the supply chain (Gutfleisch et al. 2011). This includes investing in technologies that can promote energy efficiency, utilizing alternative energy sources, and optimizing warehousing and transportation.

5. Collaboration and Information Sharing: Collaboration and integration among supply chain partners and effective information sharing are critical for implementing GSCM. This can be achieved only if trust is developed along the supply chain (Mei et al. 2023). Organizations will have to work towards establishing transparent communication channels in both directions to share relevant and real-time data, share best practices, and jointly work at achieving innovative solutions.

6. Life Cycle Thinking: GSCM incorporates a life cycle perspective from the environmental standpoint meaning that it has to be considered from product inception to disposal. This principle implies that the concept of sustainability should be embedded across the entire life cycle, from designing the product, raw material extraction to manufacturing, distribution of the products, product use, and product end-of-life management (Akter, & Saif Abu 2022). By understanding the environmental implications across each and every stage, organizations can make well-informed decisions to minimize their negative impact on the environment.

7. Continuous Improvement: Continuous improvement is a core lean principle for organizations aiming to implement GSCM. The concept adopted from Kaizen in lean management involves setting small but continuous sustainability goals, measuring performance regularly, and striving for continual improvements over the years (Menon & Ravi, 2022). Organizations should plan for small increments rather than aiming for big leaps in achieving sustainability and identify areas for improvement, and implement strategies to enhance sustainability continuously.

By adhering to these seven key principles, organizations can establish a robust foundation for effective GSCM. These principles encompass compliance with environmental standards/regulations, promoting sustainable sourcing, efficient waste reduction and recycling, working towards energy efficiency, collaboration with supply chain partners, life cycle thinking, and continuous improvement, allowing businesses to achieve their sustainability objectives.

6.5   RECENT TRENDS IN GREEN SUPPLY CHAIN MANAGEMENT

With a constant evolution in the concept of GSCM there have been various recent trends and some of them are as follows

1. Auditing suppliers from an Environmental perspective: Conducting audits both scheduled and random audits of suppliers and their manufacturing premises to assess their adherence to practicing environmentally sustainable practices and ensure they meet specific environmental standards and regulations as agreed upon (Gedam et al. 2021).

2. Collaborative recycling with partners: Collaborating with partners apart from information sharing will also have to collaborate to target recycling initiatives, promoting the recycling of materials and reducing waste generation throughout the supply chain (Gao et al. 2021).

3. Environmental risk management system: The implementation of a comprehensive risk management framework is imperative for the identification, assessment, and mitigation of environmental risks within the supply chain. It is crucial for organizations to adopt a proactive approach, ensuring the establishment of robust measures aimed at minimizing any potential adverse effects on the environment (De Oliveira, Leiras & Ceryno, 2019).

4. Environmental management systems: The implementation of environmental management systems, in accordance with the ISO 14001 standard, offers organizations a complete and comprehensive framework to effectively manage their environmental regulations and requirements and enhance their overall performance from the environmental perspective (Das et al. 2023).

5. Environmental policy for GSCM: The formulation of an environmental policy is imperative to articulate the organization's dedicated efforts to the effective implementation of GSCM practises. This environmental policy shall enlist a comprehensive framework encompassing the setting of goals, strategies to be adopted, and guidelines aimed at mitigating the adverse environmental effects associated with the organizations operations (Khattab et al. 2022).

6. Environmental training and education for partners: The implementation of sustainable practices greatly depends on the providing necessary training and educational initiatives at the organizational level as well as to supply chain partners. These programmes are to be designed to augment their comprehension of pertinent environmental concerns, sustainable best practices, and the crucial role they play in bolstering the efficacy of GSCM (Jayant & Neeru, 2020).

7. Establishing environmental policies for purchasing items: In order to pursue the sustainability of the procurement processes, it is imperative to establish well-defined environmental parameters for the selection of vendors and purchase of items. This ensures purchasing items from suppliers who exhibit robust environmental consciousness (Alshourah et al. 2022).

8. Green Inventory Management: Focus is also to be aimed to minimize surplus inventory, optimize the storage facility utilization, and thereby ensure that inventory management practises are in sync with the sustainability objectives of the organization (Gunawardena, 2021).

9. Green manufacturing: The implementation of eco-friendly manufacturing processes is of great significance in the pursuit of sustainable and environmental practices (Afum et al. 2020). Green manufacturing is achieved by adopting many practices such as minimizing resource usage, eliminating waste and reducing emissions. Firms by adopting green manufacturing can significantly contribute to the conservation and protection of our natural resources.

10. Green packaging: An extension to green manufacturing is green packaging where emphasis is laid on using ecologically sustainable materials and adopting cutting-edge designs and innovations in packaging

thereby mitigating packaging waste, nurturing recyclability, and delving into alternatives for packaging that yield minimal ecological damages (Karlsson & Karlsson, 2020).

11. Green product design: The integration of sustainability practices starts from the design of products themselves, which primarily focuses on various dimensions such as energy efficiency, reusability, recyclability, usage of environmentally sustainable raw materials, and minimizing the overall carbon footprint throughout the entire life cycle of the product (Ye, Lau & Teo, 2023).

12. Green product use: Organizations should also keep in mind the other critical stakeholders, namely the customer, in their efforts to move towards sustainability. It becomes imperative for organizations to promote and educate conscientious utilization of products among customers (Migdadi, 2019). This can be achieved by offering incentives, issuing comprehensive guidelines, spelling out clear instructions, and actively promoting rational disposal methods, energy conservation and responsible usage of the products.

13. Green retailing: The adoption of green and sustainable practises in the context of retail operations is gaining traction and it encompasses a range of wide strategies to be adopted by organizations (Adhikari, Biswas & Avittathur, 2019). These include the establishment of energy-efficient stores, the adoption of technologies to reduce carbon footprint in their retail stores, adoption of sustainable sourcing methods, and the reduction of waste.

14. Green Supplier: The strategic process of supplier selection and collaboration entails the careful evaluation and engagement of suppliers who exhibit commendable environmental performance, robust sustainability initiatives, and unwavering commitment to upholding stringent environmental standards (Ghosh, Mandal & Ray, 2022).

15. Information collaboration: Facilitating the dissemination of pertinent environmental information and data among supply chain partners in order to cultivate a culture of transparency, engender collaborative efforts, and collectively strive towards enhancing environmental performance (Mei et al. 2023).

16. Pursuing green purchasing: Green purchasing can be achieved by proactively engaging with suppliers and purchasing products that adhere to green principles, while simultaneously taking into account environmental considerations and educating the suppliers to take the sustainable journey (Tanuwijaya, Tarigan & Siagian, 2021).

17. Regular Suppliers meeting: Organizations should ideally conduct periodic meetings with suppliers to audit and review the environmental performance and metrics, and discuss the deviations. The meetings also act as a platform to share best practices and knowledge transfer and discuss the challenges faced by the suppliers and ways to overcome them (Yu, Zhang & Huo, 2021).

18. Requesting compliance statement: Firms should establish a stringent protocol mandating suppliers to furnish statements or documentation that substantiate their adherence to prevailing environmental regulations and standards that were mutually agreed upon. This can be validated by the conduct of random audits and scheduled audits (Maqbool, 2019).

19. Research collaboration with suppliers: Firms should engage with their supply partners in collaborative endeavors to spearhead research and development initiatives focussing on the creation of ecologically sustainable products, materials, or processes (Juliandina, Chelliah & Yin, 2022).

20. Total quality management with an emphasis on the environment: The adoption of Total quality management with specific emphasis on environmental concerns as a core belief is of paramount importance. The primary objective is to accomplish a consistent level of product quality while simultaneously taking care of the environment. This will help in creating a balance between offering quality to all concerned stakeholders while also minimizing environmental damage (Al-khawaldah et al. 2022).

The snapshot of the recent trends in GSCM has been represented in Table 6.2.

TABLE 6.2
Recent Trends in GSCM

	Approach	Representative References
1	Auditing suppliers from an environmental perspective	Gedam et al. 2021
2	Collaborative recycling with partners	Gao et al. 2021
3	Develop an environmental risk management system	De Oliveira,Leiras & Ceryno, 2019
4	Environmental management systems - ISO 14001 certification	Das et al. 2023
5	Environmental policy for GSCM	Khattab et al. 2022
6	Environmental training and education for partners	Jayant & Neeru, 2020
7	Establishing environmental requirements for purchasing items	Alshourah et al. 2022
8	Green inventory management	Gunawardena, 2021
9	Green manufacturing	Afum et al. 2020
10	Green packaging	Karlsson & Karlsson, 2020
11	Green product design	Ye, Lau & Teo, 2023
12	Green product use	Migdadi, 2019
13	Green retailing	Adhikari, Biswas & Avittathur, 2019
14	Green supplier	Ghosh, Mandal & Ray, 2022
15	Information collaboration	Mei et al. 2023
16	Pursuing green purchasing	Tanuwijaya,Tarigan & Siagian, 2021
17	Regular suppliers meeting	Yu, Zhang & Huo, 2021
18	Requesting compliance statement	Maqbool, 2019
19	Research & development collaboration with suppliers	Juliandina, Chelliah & Yin, 2022
20	Total quality management with emphasis on the environment	Al-khawaldah et al. 2022

Source: Author's Own

6.6 BEST PRACTICES OF GSCM

Based on the literature review the best practices that companies have adopted to attain improvements in the GSCM practices can be classified as shown in Figure 6.7.

The first classification revolves around the exploration and implementation of responsible sourcing practises as shown in Table 6.3. These practises encompass a wide range of activities, such as engaging in collaborative efforts with suppliers to procure sustainable materials, embracing renewable energy sources, optimizing packaging methods to minimizing environmental repercussions, and incorporating sustainability criteria into the supplier selection process.

The second classification places a strong emphasis on implementing methodologies that are geared towards the minimization of energy consumption, reduction of waste generation, and optimization of resource utilization as stated in Table 6.4. The comprehensive approach encompasses the integration of energy-efficient technologies, the implementation of recycling and waste reduction initiatives, as well as the adoption of water conservation measures.

The third classification aims to address and mitigate the environmental impact reduction in the activities that are associated with the supply chain of the firm as indicated in Table 6.5. These initiatives are targeted at quantifying, mitigating environmental risks and compensating carbon emissions, shifting to sustainable modes of transportation, and implementing sustainable practises within the realm of distribution and logistics.

The fourth classification focuses on sustainable practices that involve actively engaging stakeholders such as educating and raising awareness among customers, suppliers and employees, and communicating the green initiatives adopted by the firm to the suppliers, customers and stakeholders as represented in Table 6.6. It reinforces the importance of supply chain visibility and supply chain transparency

FIGURE 6.7   Best practices of GSCM.

TABLE 6.3

Best Practices of GSCM in Sustainable Sourcing and Material Management

Best Practice	Description	Company	References
Sustainable Sourcing and Material Management:			
Supplier collaboration for sustainable sourcing	Collaborating with suppliers to ensure suppliers practice sustainable initiatives, including usage of environmentally friendly materials, ethical labor practices, ethical wage rates etc.	Nike	Hamner, B. (2006)
Adoption of renewable energy sources	Implementing renewable energy alternatives such as shifting from fossil fuels to sustainable power sources such as solar or wind power effectively decrease greenhouse gas emissions.	Starbucks	Wang, Dargusch & Hill, 2022
Green packaging and materials optimization	Adopting eco-friendly packaging materials, eliminating or reducing packaging waste, and working towards innovative packaging designs ensuring reduced environmental impact.	Toyota Motors	Gutfleisch et al. 2011
Integration of sustainability as a key supplier selection criterion	Inculcating sustainability as a criterion in supplier selection considering factors such as environmental performance or social responsibility etc.	Honda Motors	Ramadhanti & Pulansari, 2022

Source: Author's Own

and building long-term relationships with stakeholders to drive sustainability efforts across the supply chain.

The fifth classification revolves around establishing a robust and reliable measurement system, effective tracking and reporting of sustainability performance, and adopting various digital technologies for real-time monitoring of environmental activities and their outcomes as stated in Table 6.7. It highlights the importance of data-driven sustainability improvements and works towards continuous improvement.

The sixth and last classification focuses on practices that are associated with circular economy principles, including lifecycle assessments, implementing efficient reverse logistics for product returns at the end of their life cycle, recycling, and designing products that have lesser impact on the environment as listed in Table 6.8. It aims to adopt innovative methods and approaches to minimize waste and optimize resource utilization.

TABLE 6.4

Best Practices of GSCM in Energy and Resource Efficiency

Best Practice	Description	Company	References
Energy and Resource Efficiency:			
Implementation of energy-efficient technologies	Adoption of energy-efficient practices and technologies to optimize energy consumption and thereby achieve overall operational efficiency.	General Motors	Kaur & Sharma, 2017
Recycling and waste reduction initiatives	Implementing efficient recycling initiatives and waste reduction approaches to minimize environmental influence	Ikea	Alrubah et al. 2020
Water conservation and efficient water management	Initiating water preservation and conservation measures and effective water management practices to reduce water consumption and reduce water-related environmental effects.	Nestle	Galli & Vousvouras, 2020

Source: Author's Own

TABLE 6.5

Best Practices of GSCM – Environmental Impact Reduction

Best Practice	Description	Company	References
Environmental Impact Reduction:			
Carbon footprint reduction strategies	Designing and adopting strategies to understand, measure, analyse, reduce, and offset carbon emissions that are related to the firms' supply chain activities.	Microsoft	Sehgal et al. 2020
Adoption of sustainable transportation methods	Shifting to a more sustainable transportation mode, such as rail or electric vehicles, from traditional transportation modes to minimize carbon emissions and reduce the impact on the environment.	Ikea	Laurin & Fantazy, 2017
Green logistics and distribution	Adopting environmentally friendly practices in logistics and distribution services, by using fuel-efficient vehicles including route optimization, and emissions reduction strategies.	DHL	Dhir et al. 2019

Source: Author's Own

TABLE 6.6

Best Practices of GSCM – Stakeholder Engagement and Transparency

Best Practice	Description	Company	References
Stakeholder Engagement and Transparency:			
Stakeholder engagement and transparency	Engaging with stakeholders, including customers, employees, and communities, and promoting transparency in environmental performance and sustainability initiatives.	Apple	Lu et al. 2023
Employee training and awareness programs	Providing training and awareness programs to employees to promote environmentally responsible practices and enhance their understanding of sustainability.	Toyota	Benkarim & Imbeau, 2022
Green marketing and communication strategies	Promoting and communicating the organization's green initiatives and sustainable products to customers and stakeholders to drive awareness and demand for eco-friendly options.	Wipro	Rani, 2022

Source: Author's Own

TABLE 6.7

Best Practices of GSCM – Performance Measurement and Reporting

Best Practice	Description	Company	References
Performance Measurement and Reporting:			
Measurement and reporting of sustainability performance	Establishing robust measurement systems to track and report on sustainability performance, including key environmental metrics and progress towards sustainability goals.	Schneider Electric	Robert, Giuliani & Gurau, 2022
Use of digital technologies for environmental monitoring	Leveraging digital technologies, such as IoT sensors or data analytics, to monitor and optimize environmental performance, resource consumption, and emissions.	Rolls Royce	Rusch, Schöggl & Baumgartner, 2023

Source: Author's Own

TABLE 6.8

Best Practices of GSCM – Circular Economy and Innovation

Best Practice	Description	Company	References
Circular Economy and Innovation:			
Lifecycle assessment and eco-design of products	Conducting lifecycle assessments of products to identify and minimize their environmental impacts, and incorporating eco-design principles in product development.	L'Oreal	Gatt & Refalo, 2022
Reverse logistics for product returns and recycling	Establishing systems for managing product returns, refurbishing or recycling returned products, and reducing waste generated in the reverse supply chain.	Apple Inc.	Wilson & Goffnett, 2022
Incorporation of circular economy principles	Embracing circular economy principles by designing products for reuse, recycling, or remanufacturing, and optimizing resource utilization throughout the supply chain.	Unilever	Rizkovic, 2023

Source: Author's Own

6.6.1   FRAMEWORK FOR IMPLEMENTING GREEN SUPPLY CHAIN MANAGEMENT

In spite of the numerous benefits that GSCM offers, organizations have been facing challenges in Implementing GSCM. Adoption of GSCM requires a committed and systematic approach that assimilates sustainability principles and practices deep into the organization's supply chain network. The following is the step-by-step framework for organizations to implement GSCM as shown in Figure 6.8:

1. Establish Clear Environmental Objectives:
   - Define explicit environmental objectives that are aligned with the organization's goals.
   - Identify and list key performance indicators (KPIs) to measure and analyse the progress of the environmental initiatives.
2. Map the Current Supply Chain Practices:
   - Perform a comprehensive assessment of the organization's existing supply chain processes to assess the current environmental impacts, risks, and opportunities.
   - Identify and measure the current carbon footprint, energy usage, waste generation, and natural resources usage across the supply chain.
3. Involve Suppliers and Stakeholders:
   - Involve and collaborate with suppliers, customers and other stake-holders to spread the organization's firm commitment to practice GSCM.

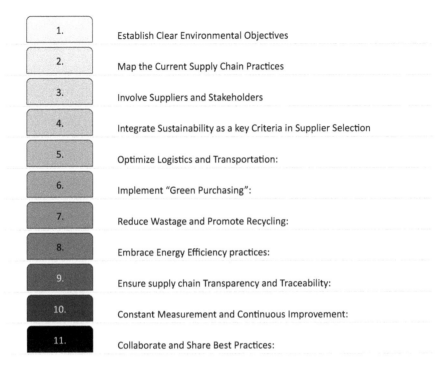

1. Establish Clear Environmental Objectives

2. Map the Current Supply Chain Practices

3. Involve Suppliers and Stakeholders

4. Integrate Sustainability as a key Criteria in Supplier Selection

5. Optimize Logistics and Transportation:

6. Implement "Green Purchasing":

7. Reduce Wastage and Promote Recycling:

8. Embrace Energy Efficiency practices:

9. Ensure supply chain Transparency and Traceability:

10. Constant Measurement and Continuous Improvement:

11. Collaborate and Share Best Practices:

FIGURE 6.8   Framework for implementing green supply chain management.

Source: Author's Own.

- Foster long-term partnerships and discussions to motivate and encourage the adoption of sustainable practices throughout the entire supply chain.
4. Integrate Sustainability as a key Criteria in Supplier Selection:
    - Include firm guidelines and specific criteria for selecting suppliers also considering their environmental performance and sustainability practices.
    - Support and acknowledge suppliers who adhere and follow green standards and demonstrate a focussed commitment towards sustainability.
5. Optimize Logistics and Transportation:
    - Do a mapping of the existing transportation modes and travel routes and modes to identify opportunities for optimizing transportation, reducing carbon emissions and improving sustainability.
    - Contemplate identifying alternative transportation methods, such as electric vehicles or public transportation such as railways to minimize environmental impact.
6. Implement "Green Purchasing":
    - Develop and implement "green Purchasing" that promotes the purchase of materials from suppliers who adopt environmentally friendly practices.
    - Consider factors apart from normal parameters such as cost, quality, delivery, flexibility and move towards parameters such as

recyclability, usage of renewable resources, and efficient waste management practices while making purchase decisions.

7. Reduce Wastage and Promote Recycling:
   * Identify, analyse and implement waste reduction strategies across the supply chain and adopt and promote recycling programs within the firm level as well as at the supplier's level.
   * Identify strategies to minimize packaging waste, advocate the reuse of packaging materials, and encourage supply chain partners to adopt sustainable packaging practices.

8. Embrace Energy Efficiency practices:
   * Adopt energy-efficient practices and the latest technologies and practices that promote energy efficiency across the supply chain, including their facility, warehouses, transportation and operational activities.
   * Implement efficient energy management systems, monitor and progress energy usage, and explore renewable energy alternatives.

9. Ensure supply chain Transparency and Traceability:
   * To have effective GSCM practices firms should first establish supply chain transparency and traceability across the supply chain.
   * Establish transparency and traceability systems to ensure traceability and transparency in sourcing, promoting sustainable practices.

10. Constant Measurement and Continuous Improvement:
    * Make efforts to regularly measure and evaluate the progress in the sustainable measures against the predefined KPIs.
    * Identify a team to constantly Identify areas for improvement in the sustainable space and set targets for improvements in the environmental areas.
    * Encouraging employee engagement and innovation and thereby creating a culture of continuous improvement in the supply chain network.

11. Collaborate and Share Best Practices:
    * Engage with all stakeholders such as suppliers, customers, industry associations, Government bodies, and other organizations to share best practices of sustainability and learn from each other.
    * Collaborate jointly on research and innovation progress in the field of GSCM collectively.

6.7   CONCLUSION

"Uncovering the Shifting Landscape: A Comprehensive Analysis of Emerging Trends and Best Practices Adopted in Green Supply Chain Management" offers a comprehensive review of the progressive shift witnessed in individuals, corporations and governments across the globe, as they strive to align supply chain methodologies and practices with the imperatives of environmental sustainability. This chapter has undertaken a comprehensive exploration of the understanding, of the recent trends and best practices that have been adopted in the domain of green supply chain management, thereby furnishing invaluable insights and a blueprint for enterprises to effectively incorporate sustainable practises.

The chapter begins by establishing the mounting emphasis on green supply chain management which is driven by mounting Governmental environmental regulations and increasing stakeholder expectations towards a sustainable future. The next part of the chapter elaborates on a detailed literature review of the topic and aims to give the reader a detailed analysis of the field of GSCM. The chapter aims to provide significant insight into the various best practices adopted by specific organizations in their pursuit of adopting green supply chains. The chapter also aims to present a step-by-step detailed framework for entities and organizations to implement sustainable practices within their firm.

This chapter contributes novelty to the vast existing literature by presenting an in-depth analysis of the evolving field of green supply chain management. It provides significant insights into the various recent good practices adopted by organizations, which can serve as an ideal starting point for organizations that aim to adopt GSCM practices into their operations. It also reiterates the role of recent and emerging industry 4.0 technologies in promoting sustainability across supply chains. This book chapter serves as a unique and valuable resource for Policymakers, academicians, researchers, organizations and practitioners interested in GSCM to have a clear understanding and navigate the shifting landscape of green supply chain management. It offers significant insights into recent and emerging trends as well as various sustainable best practices in the supply chain context that are adopted by organizations.

In conclusion, "Uncovering the Shifting Landscape: A Comprehensive Analysis of Emerging Trends and Best Practices Adopted in Green Supply Chain Management" provides a comprehensive overview of the evolving field of GSCM. The chapter offers significant and valuable insights into recent trends and best practices, and provides a framework organization to implement GSCM. By highlighting the role of emerging technologies and presenting a step-by-step framework for organizations to go for implementation, this chapter contributes to the understanding and adoption of sustainable practices in supply chain management.

REFERENCES

Aćimović, S., Mijušković, V., & Rajić, V. (2020). The impact of reverse logistics onto green supply chain competitiveness evidence from Serbian consumers. [Reverse logistics in green supply chain] *International Journal of Retail & Distribution Management, 48*(9), 1003–1021. 10.1108/IJRDM-04-2019-0142

Adhikari, A., Biswas, I., & Avittathur, B. (2019). Green retailing: A new paradigm in supply chain management. In *Green business: Concepts, methodologies, tools, and applications* (pp. 1489–1508). Pennsylvania, US: IGI Global.

Afum, E., Osei-Ahenkan, V. Y., Agyabeng-Mensah, Y., Owusu, J. A., Kusi, L. Y., & Ankomah, J. (2020). Green manufacturing practices and sustainable performance among Ghanaian manufacturing SMEs: The explanatory link of green supply chain integration. *Management of Environmental Quality: An International Journal, 31*(6), 1457–1475.

Aityassine, F., Aldiabat, B., Al-rjoub, S., Aldaihani, F., Al-Shorman, H., & Al-Hawary, S. (2021). The mediating effect of just in time on the relationship between green supply chain management practices and performance in the manufacturing companies. *Uncertain Supply Chain Management*, *9*(4), 1081–1090.

Akter, R. R., & Saif Abu, N. M. (2022). Impact of green supply chain management (GSCM) on business performance and environmental sustainability: Case of a developing country. *Business Perspectives and Research*, *10*(1), 140–163. 10.1177/2278533720983089

AlBrakat, N., Al-Hawary, S., & Muflih, S. (2023). Green supply chain practices and their effects on operational performance: An experimental study in Jordanian private hospitals. *Uncertain Supply Chain Management*, *11*(2), 523–532.

Al-khawaldah, R., Al-zoubi, W., Alshaer, S., Almarshad, M., ALShalabi, F., Altahrawi, M., & Al-Hawary, S. (2022). Green supply chain management and competitive advantage: The mediating role of organizational ambidexterity. *Uncertain Supply Chain Management*, *10*(3), 961–972.

Alrubah, S. A., Alsubaie, L. K., Quttainah, M. A., Pal, M., Pandey, R., Kee, D. M. H., … & Aishan, N. (2020). Factors affecting environmental performance: A study of IKEA. *International Journal of Tourism and Hospitality in Asia Pacific (IJTHAP)*, *3*(3), 79–89.

Alshourah, S., Mansour, M., AlZeaideen, K., & Azzam, Z. (2022). The effect of top management support and support supplier development on green supply chain management in the construction Jordanian. In *The implementation of smart technologies for business success and sustainability: During COVID-19 crises in developing countries* (pp. 191–202). Cham: Springer International Publishing.

Awan, F. H., Dunnan, L., Jamil, K., Mustafa, S., Atif, M., Gul, R. F., & Guangyu, Q. (2022). Mediating role of green supply chain management between lean manufacturing practices and sustainable performance. *Frontiers in Psychology*, *12*, 810504.

Benkarim, A., & Imbeau, D. (2022). Investigating the implementation of Toyota's human resources management practices in the aerospace industry. *Merits*, *2*(3), 126–145.

Chkanikova, O., & Sroufe, R. (2021). Third-party sustainability certifications in food retailing: Certification design from a sustainable supply chain management perspective. *Journal of Cleaner Production*, *282*, 124344.

Das, G., Li, S., Tunio, R. A., Jamali, R. H., Ullah, I., & Fernando, K. W. T. M. (2023). The implementation of green supply chain management (GSCM) and environmental management system (EMS) practices and its impact on market competitiveness during COVID-19. *Environmental Science and Pollution Research*, 1–16.

De Oliveira, F. N., Leiras, A., & Ceryno, P. (2019). Environmental risk management in supply chains: A taxonomy, a framework and future research avenues. *Journal of Cleaner Production*, *232*, 1257–1271.

Despoudi, S. (2020). Green supply chain. In *The interaction of food industry and environment* (pp. 35–61). Massachusetts, US: Academic Press.

Dheeraj, N., & Vishal, N. (1992). An overview of green supply chain management in India. *Research Journal of Recent Sciences ISSN*, *2277*, 2502.

Dhir, S., Sushil, Dhir, S., & Sushil. (2019). DHL. Cases in strategic management: A flexibility perspective, 55–72.

Einizadeh, A., & Kasraei, A. (2021). Proposing a model of green supply chain management based on new product development (NPD) in auto industry. *Journal of Economics & Management Research*, *10*.

Galli, C. C., & Vousvouras, C. (2020). Nestlé caring for water. *International Journal of Water Resources Development*, *36*(6), 1093–1104.

Gao, S., Qiao, R., Lim, M. K., Li, C., Qu, Y., & Xia, L. (2021). Integrating corporate website information into qualitative assessment for benchmarking green supply chain management practices for the chemical industry. *Journal of Cleaner Production*, *311*, 127590.

García Alcaraz, J. L., Díaz Reza, J. R., Arredondo Soto, K. C., Hernández Escobedo, G., Happonen, A., Puig, I., Vidal, R., & Jiménez Macías, E. (2022). Effect of green supply chain management practices on environmental performance: Case of Mexican manufacturing companies. *Mathematics*, *10*(11), 1877.

Gatt, I. J., & Refalo, P. (2022). Reusability and recyclability of plastic cosmetic packaging: A life cycle assessment. *Resources, Conservation & Recycling Advances*, *15*, 200098.

Gedam, V. V., Raut, R. D., Lopes de Sousa Jabbour, A. B., Narkhede, B. E., & Grebinevych, O. (2021). Sustainable manufacturing and green human resources: Critical success factors in the automotive sector. *Business Strategy and the Environment*, *30*(2), 1296–1313.

Gelmez, E. (2020). The mediation role of environmental performance in the effects of green supply chain management practices on business performance. *Avrupa Bilim ve Teknoloji Dergisi*, (19), 606–613.

Ghosh, S., Mandal, M. C., & Ray, A. (2022). Green supply chain management framework for supplier selection: An integrated multi-criteria decision-making approach. *International Journal of Management Science and Engineering Management*, *17*(3), 205–219.

Gunawardena, L. M. (2021). *Impact of Green Supply Chain Management Practices on Cost Efficiency: Supermarket Chain in Western Province-Sri Lanka* (Doctoral dissertation).

Gutfleisch, O., Willard, M. A., Brück, E., Chen, C. H., Sankar, S. G., & Liu, J. P. (2011). Magnetic materials and devices for the 21st century: Stronger, lighter, and more energy efficient. *Advanced materials*, *23*(7), 821–842.

Hamner, B. (2006). Effects of green purchasing strategies on supplier behaviour. In *Greening the supply chain* (pp. 25–37). London: Springer London.

Jayant, A., & Neeru. (2020). Decision support framework for smart implementation of green supply chain management practices. *New Paradigm of Industry 4.0: Internet of Things, Big Data & Cyber Physical Systems*, 49–98.

Juliandina, M., Chelliah, S., & Yin, T. S. (2022). Green supply chain management (gscm) affecting an organization's sustainability performance in indonesia wooden furniture industry. *International Journal of Accounting*, *7*(45), 11–41.

Kale, P. B. (2021). An analysis of evolution from traditional to green supply chain. *International Journal*, *6*(1), 12–21.

Karabacak, Z., & Saygili, M. S. (2022). Green practices in supply chain management: Case studies. *Journal of Business and Trade*, *3*(1), 65–81.

Karlsson, E., & Karlsson, A. (2020). *Green supply chain practices for a consumer health business in the UK market – The implications of implementing Green Packaging*. Gothenburg, Sweden.

Kaur, A., & Sharma, P. C. (2017). Sustainability as a strategy incorporated in decision making at supply chain management – Case study of General Motors. *International Journal of Sustainable Strategic Management*, *5*(3), 183–200.

Khan, M. T., Idrees, M. D., Rauf, M., Sami, A., Ansari, A., & Jamil, A. (2022). Green supply chain management practices' impact on operational performance with the mediation of technological innovation. *Sustainability*, *14*(6), 3362.

Khattab, S., Shaar, I., Alkaied, R., & Qutaishat, F. (2022). The relationship between big data analytics and green supply chain management by looking at the role of environmental orientation: Evidence from emerging economy. *Uncertain Supply Chain Management*, *10*(2), 303–314.

Krishnan, R., Agarwal, R., Bajada, C., & Arshinder, K. (2020). Redesigning a food supply chain for environmental sustainability–An analysis of resource use and recovery. *Journal of Cleaner Production*, *242*, 118374.

Kuiti, M. R., Ghosh, D., Gouda, S., Swami, S., & Shankar, R. (2019). Integrated product design, shelf-space allocation and transportation decisions in green supply chains. *International Journal of Production Research*, *57*(19), 6181–6201.

Laguir, I., Stekelorum, R., & El Baz, J. (2021). Going green? Investigating the relationships between proactive environmental strategy, GSCM practices and performances of third-party logistics providers (TPLs). *Production Planning & Control, 32*(13), 1049–1062.

Laurin, F., & Fantazy, K. (2017). Sustainable supply chain management: A case study at IKEA. *Transnational Corporations Review, 9*(4), 309–318.

Lee, S. Y. (2015). The effects of green supply chain management on the supplier's performance through social capital accumulation. *Supply Chain Management: An International Journal, 20*(1), 42–55.

Leng, J., Ruan, G., Jiang, P., Xu, K., Liu, Q., Zhou, X., & Liu, C. (2020). Blockchain-empowered sustainable manufacturing and product lifecycle management in industry 4.0: A survey. *Renewable and sustainable energy reviews, 132*, 110112.

Liu, J., Zhu, Q., & Feng, Y. (2019). The circular economy and green supply chain management. *Handbook on the Sustainable Supply Chain, 27*.

Liu, J., Chen, M., & Liu, H. (2020). The role of big data analytics in enabling green supply chain management: A literature review. *Journal of Data, Information and Management, 2*, 75–83.

Lu, W. M., Kweh, Q. L., Ting, I. W. K., & Ren, C. (2023). How does stakeholder engagement through environmental, social, and governance affect eco-efficiency and profitability efficiency? Zooming into Apple Inc.'s counterparts. *Business Strategy and the Environment, 32*(1), 587–601.

Maqbool, A. (2019, June). Modeling the barriers of sustainable supply chain practices: A Pakistani perspective. In *Proceedings of the Thirteenth International Conference on Management Science and Engineering Management: Volume 2* (Vol. 1002, p. 348). Cham, Switzerland: Springer.

Mathu, K. (2019). Green supply chain management: A precursor to green purchasing. *Green Practices and Strategies in Supply Chain Management, 43*.

Mei, T., Qin, Y., Li, P., & Deng, Y. (2023). Influence mechanism of construction supply chain information collaboration based on structural equation model. *Sustainability, 15*(3), 2155.

Menon, R. R., & Ravi, V. (2022). An analysis of barriers affecting implementation of sustainable supply chain management in electronics industry: a Grey-DEMATEL approach. *Journal of Modelling in Management, 17*(4), 1319–1350.

Migdadi, Y. K. A. A. (2019). The effective practices of mobile phone producers' green supply chain management in reducing GHG emissions. *Environmental Quality Management, 28*(3), 17–32.

Mitra, S., & Datta, P. P. (2014). Adoption of green supply chain management practices and their impact on performance: an exploratory study of Indian manufacturing firms. *International Journal of Production Research, 52*(7), 2085–2107.

Nyachomba, J., & Achuora, J. (2022). Determinants of green supply chain management practices on performance of gas manufacturing firms In Kenya. *International Journal of Supply Chain and Logistics, 6*(1), 33–57.

Park, S. R., Kim, S. T., & Lee, H. H. (2022). Green supply chain management efforts of first-tier suppliers on economic and business performances in the electronics industry. *Sustainability, 14*(3), 1836.

Plaza-Úbeda, J. A., Abad-Segura, E., de Burgos-Jiménez, J., Boteva-Asenova, A., & Belmonte-Ureña, L. J. (2020). Trends and new challenges in the green supply chain: The reverse logistics. *Sustainability, 13*(1), 331.

Rajabion, L., Khorraminia, M., Andjomshoaa, A., Ghafouri-Azar, M., & Molavi, H. (2019). A new model for assessing the impact of the urban intelligent transportation system, farmers' knowledge and business processes on the success of green supply chain management system for urban distribution of agricultural products. *Journal of Retailing and Consumer Services, 50*, 154–162.

Ramadhanti, V. I., & Pulansari, F. (2022). Integration of fuzzy AHP and fuzzy TOPSIS for green supplier selection of mindi wood raw materials. *Jurnal sistem dan manajemen industri, 6*(1), 1–13.

Rani, N. (2022). Green marketing: Opportunities and challenges. *New Horizons In Business World, 335.*

Reche, A. Y. U., Junior, O. C., Estorilio, C. C. A., & Rudek, M. (2020). Integrated product development process and green supply chain management: Contributions, limitations and applications. *Journal of Cleaner Production, 249,* 119429.

Rizkovic, A. (2023). Plastic Packaging Waste Management by L'Oreal and Unilever: A Circular Economy Perspective. *Journal of World Trade Studies, 7*(1), 60–75.

Robert, M., Giuliani, P., & Gurau, C. (2022). Implementing industry 4.0 real-time performance management systems: The case of Schneider Electric. *Production Planning & Control, 33*(2–3), 244–260.

Rusch, M., Schöggl, J. P., & Baumgartner, R. J. (2023). Application of digital technologies for sustainable product management in a circular economy: A review. *Business Strategy and the Environment, 32*(3), 1159–1174.

Sarkis, J., Kouhizadeh, M., & Zhu, Q. S. (2021). Digitalization and the greening of supply chains. *Industrial Management & Data Systems, 121*(1), 65–85.

Samekto, A. A., & Kristiyanti, M. (2022). Green supply chains and sustainable maritime transportation in Covid-19 pandemic. *International Journal of Mechanical Engineering, 7,* 1512–1517.

Sant, T. G. (2022). Distribution channel coordination in green supply chain management in the presence of price premium effects. *International Journal of Services and Operations Management, 41*(1–2), 142–162.

Saridogan, M. (2012). The impact of green supply chain management on transportation cost reduction in Turkey. *International Review of Management and Marketing, 2*(2), 112–121.

Sehgal, G., Kee, D. M. H., Low, A. R., Chin, Y. S., Woo, E. M. Y., Lee, P. F., & Almutairi, F. (2020). Corporate social responsibility: A case study of Microsoft Corporation. *Asia Pacific Journal of Management and Education (APJME), 3*(1), 63–71.

Sharma, R., Rana, G., & Agarwal, S. (Eds.). (2022). *Entrepreneurial innovations, models, and implementation strategies for Industry 4.0* (1st ed.). Boca Raton: CRC Press. 10.1201/9781003217084

Shekari, H., Shirazi, S., Afshari, M., & Veyseh, S. (2011). Analyzing the key factors affecting the green supply chain management: A case study of steel industry. *Management Science Letters, 1*(4), 541–550.

Sheng, X., Chen, L., Yuan, X., Tang, Y., Yuan, Q., Chen, R., ... & Liu, H. (2023). Green supply chain management for a more sustainable manufacturing industry in China: A critical review. *Environment, Development and Sustainability, 25*(2), 1151–1183.

Soyege, O. O., Makinde, G. O., & Akinlabi, B. H. (2023). Green supply chain management and organizational performance of fast-moving consumer goods firms in Lagos Nigeria. *International Journal of Entrepreneurship, 6*(2), 1–20.

Srivastava, S. K. (2007). Green supply-chain management: A state-of-the-art literature review. *International Journal of Management Reviews, 9*(1), 53–80.

Stekelorum, R., Laguir, I., Gupta, S., & Kumar, S. (2021). Green supply chain management practices and third-party logistics providers' performances: A fuzzy-set approach. *International Journal of Production Economics, 235,* 108093.

Sugandini, D., Susilowati, C., Siswanti, Y., & Syafri, W. (2020). Green supply management and green marketing strategy on green purchase intention: SMEs cases. *Journal of Industrial Engineering and Management (JIEM), 13*(1), 79–92.

Sutawijaya, A. H., & Nawangsari, L. C. (2020). What is the impact of industry 4.0 to green supply chain. *Journal of Environmental Treatment Techniques, 8,* 207–213.

Tanuwijaya, N. C., Tarigan, Z. J. H., & Siagian, H. (2021). The effect of top management commitment on firm performance through the green purchasing and supplier relationship management in 3-star hotel industry in Surabaya. *Petra International Journal of Business Studies, 4*(2), 169–181.

Tseng, M. L., Islam, M. S., Karia, N., Fauzi, F. A., & Afrin, S. (2019). A literature review on green supply chain management: Trends and future challenges. *Resources, Conservation and Recycling, 141,* 145–162.

Uemura Reche, A. Y., Canciglieri Junior, O., Szejka, A. L., & Rudek, M. (2022). Proposal for a preliminary model of integrated product development process oriented by green supply chain management. *Sustainability, 14*(4), 2190.

Venhoeven, L. A., Bolderdijk, J. W., & Steg, L. (2020). Why going green feels good. *Journal of Environmental Psychology, 71,* 101492.

Wang, Y., Dargusch, P., & Hill, G. (2022). How do World-renowned Coffee Companies Manage Carbon Emissions? A Case Study of Starbucks. *Advances in Environmental and Engineering Research, 3*(2), 1–13.

Wassie, S. B. (2020). Natural resource degradation tendencies in Ethiopia: A review. *Environmental Systems Research, 9*(1), 1–29.

Wei, G. (2023). The Impact of Blockchain Technology on Integrated Green Supply Chain Management in China: A Conceptual Study. *Journal of Digitainability, Realism & Mastery (DREAM), 2*(02), 58–65.

Wilson, M., & Goffnett, S. (2022). Reverse logistics: Understanding end-of-life product management. *Business Horizons, 65*(5), 643–655.

Xu, Y., Liu, A., Li, Z., Li, J., Xiong, J., & Fan, P. (2023). Review of Green Supply-Chain Management Diffusion in the Context of Energy Transformation. *Energies, 16*(2), 686.

Yang, C. S. (2018). An analysis of institutional pressures, green supply chain management, and green performance in the container shipping context. *Transportation Research Part D: Transport and Environment, 61,* 246–260.

Ye, Y., Lau, K. H., & Teo, L. (2023). Alignment of green supply chain strategies and operations from a product perspective. *The International Journal of Logistics Management.* https://doi.org/10.1108/IJLM-11-2021-0557

Yu, Y., Zhang, M., & Huo, B. (2021). The impact of relational capital on green supply chain management and financial performance. *Production Planning & Control, 32*(10), 861–874.

Yun, C., Shun, M., Jackson, K., Newiduom, L., & Browndi, I. (2023). Integrating life cycle assessment and green supply chain management for sustainable business practices. *International Journal of Engineering and Applied Sciences, 12*(01), 198–202.

Zhu, Q., & Sarkis, J. (2007). The moderating effects of institutional pressures on emergent green supply chain practices and performance. *International Journal of Production Research, 45*(18-19), 4333–4355.

Zsidisin, G. A., & Siferd, S. P. (2001). Environmental purchasing: A framework for theory development. *European Journal of Purchasing & Supply Management, 7*(1), 61–73.

7 Green HRM

Eco-Friendly Methods for the Advancement of Employment

Alex Khang, Geeta Rana, and Deepti Dubey

7.1 INTRODUCTION

The term "green HRM" (GHRM) is starting to gain popularity in the human resource management (HRM) industry. GHRM is an ecological practise that aids in making an organisation environmentally pleasant, essentially to keep employees aware of sustainability and get a competitive edge in the global marketplace. Organisations must adopt HRM practises that ensure a positive environmental impact in the modern world(Bhakti & Sharma, 2020). Examples of these practises include online recruitment, electronic file maintenance, online assessment and interviews, using shared cars, environmental training, and rewards based on green initiatives. Nowadays, companies are making an effort to lessen their adverse effects on the environment by raising their awareness of and expertise in these sustainable issues. Because organisations adopt new strategies and policies primarily for their own profit, it is more important to concentrate on the results of GHRM practise implementation that supports organisations (Jabbar & Abid, 2015; Rana & Arya, 2023). Given that GHRM is a relatively new area of study in organisational studies (Arulrajah, Opatha, & Nawaratne, 2015), the study concentrates on GHRM's sustainable practises in light of recent research endeavours undertaken by HRM researchers. Above all, this study aims to assist the organisation in comprehending how GHRM is implemented and how it significantly affects employment.

7.1.1 OBJECTIVES OF THE STUDY

This chapter's main goal is as follows:

- To examine GHRM practises as a whole and how they affect employment.
- Investigating the uses of GHRM is one of the other important goals.
- To study GHRM's advantages and difficulties.

DOI: 10.1201/9781003458944-7

7.2   RESEARCH METHODOLOGY

This study's entire foundation is secondary data, primarily from reviews of other researchers' works in the literature. This study's qualitative methodology is based on a thorough examination of recent publications on GHRM. Data for the study is gathered from a variety of websites, media articles, e-papers, and online journals. The primary focus for presenting the study's aims was on current research efforts.

7.3   LITERATURE REVIEW

Green laws ought to be implemented in order to guarantee that this planet is still a respectable place to live (Zubair et al., 2019). Public and commercial businesses can greatly enhance the state of the environment if they incorporate a number of essential eco-friendly practises into their operations.

For some employees, the outcomes of both green and non-green labour were better. In the meantime, improved economic and eco-performance, increased resource efficiency, the establishment of an eco-friendly workplace and organisational culture, and the development of a positive corporate image were among the benefits of implementing GHRM at the organisational level. It is anticipated that this study will broaden our understanding of the benefits that GHRM provides to businesses and how it may be applied.

Employees who receive GHRM are better equipped to understand and apply green culture in their daily lives (Muster & Schrader, 2011: Rana & Sharma 2017). Scholars such as Chams and Garca-Blandón (2019), Muisyo et al. (2021) have illustrated the importance of using GHRM practises to promote a sustainable workplace.

A study by Shafaei, Nejati, and Mohd Yusoff (2020) examines the effects of GHRM on worker productivity and welfare. Employee work satisfaction is positively impacted by GHRM, according to the study.

According to Ali et al. (2020), GHRM is a contemporary management concept designed to affect employees' environmental behaviours. Green HR advances the cause of environmental sustainability by utilising HRM practises to enhance the efficient use of resources within firm organisations.

Organisations can lessen the carbon footprint of their workforce by introducing GHRM practises, which include flexible work hours, electronic filing, job and car sharing, teleconferencing, virtual interviewing, online training, recycling, telecommuting, and energy-efficient office space. Incorporating eco-friendly HR practises and measures for sustainable resource consumption, GHRM leads to better work-related attitudes, decreased waste, increased productivity, improved work/life balance, lower costs, and improved employee performance and retention (Meily & Saragih, 2013).

HRM functions have an innate ability to help organisations and their workforce become more environmentally friendly. HRM has enormous potential to help organisations and their activities become more environmentally friendly, from job design to employee relations. Understanding the breadth and depth of GHRM in order to transform their organisations into green entities is the main challenge facing HR practitioners. It has also been noted that for GHRM practises to be implemented effectively, senior management must be very involved and supportive.

Numerous studies have been carried out with the realisation that one of the largest corporate initiatives is the greening of HRM practises. Dr Parul Deshwal's (2015) research indicates that companies are implementing GHRM as a sustainable business practise. GHRM comprises two primary components: ecologically sustainable HR practises and knowledge capital preservation, which enhances industry participants' awareness of their business and CSR obligations.

The primary focus of the bulk of studies is on finding ways to reduce an HRM department's negative impact on the environment. These days, a significant number of companies place a significant emphasis on "greening" their workforces by integrating environmentally friendly HRM encompasses several practises such as recruitment, incentives, education, and outreach. The authors Arulrajah, Opatha & Nawaratne (2015) have identified and highlighted several GHRM approaches that are found within the 12 functions of HRM: job design, job analysis, human resource planning, recruitment, selection, induction, performance evaluation, training and development, reward management, discipline management, health and safety management, and employee relations. The study's conclusions suggest that if companies guarantee green performance, green behaviours, green attitudes, and green competence, they may be able to enhance their environmental performance in a way that is more environmentally responsible.

7.3.1 GREEN HRM

Its primary goal is to make HRM practises more environmentally friendly. HRM encompasses all of its activities, from planning HR to rewarding staff members who participate in this environmental initiative. The concern of people management policies and practises with regard to the larger corporate environmental programme is sometimes referred to as "green HRM" (Deshwal, 2015). Maintaining nature-oriented HRM practises is the primary goal of GHRM in order to control the organisation's effective CSR level. Today, however, the concept of GHRM encompasses not only environmental awareness but also the organisation's and the employees' social and economic well-being in a wider sense.

7.3.2 GREEN HR PRACTICES

Adopting Green Practises makes it feasible to be affordable, practical, and ecological all at once, claims Amutha (2017). Here are a few eco-friendly tips to help you stay green.

Recycling, Green Printing, Green Manufacturing, Virtual Interviews and Teleconferences, Green Disposal of Staff ID Cards, and Virtual Interviews and Teleconferences Green HR entails lowering carbon footprints in a variety of ways, including less printing on paper, video conferencing, and interviews, amongst other things. Green HR can reduce employee carbon footprints through electronic filing. Telecommuting, online training, energy-efficient office buildings, green payroll, company transit, and electronic filing are just a few examples of green practises that businesses can implement.

For any organisation that is conscious of the need to maintain a balance between HR practises and the environment, the Human Resource Department must implement these GHRM practises. The HR department should be a key player in developing green policies, putting them into practise, and ensuring that workers have a sustainable workplace. Now that the functions of GHRM have been examined, this study attempts to assess how these practises support sustainable employment within the company by analysing a number of noteworthy recent studies.

7.4 FUNCTIONS OF GHRM

Several scholars have classified the GHRM functions, which include hiring and firing. This research article aims to analyse important recent studies in order to explain the emerging GHRM functions. The modern, automated green HRM functions that encourage employment inside a company are discussed in the next section.

7.4.1 GREEN RECRUITMENT

Hiring individuals that care about sustainable practises and are mindful of upholding a sustainable environment within the company is essentially the goal of green recruitment. It also has to do with incorporating eco-friendly hiring practises. Although the term "green recruitment" is vague, it often refers to hiring practises that reduce their negative effects on the environment by using less paper. According to Ahmad (2015), green recruiting is a strategy that prioritises and integrates the environment into the business. According to these definitions, "green recruitment" refers to the process of enticing candidates who can preserve the environment and putting in place an online application process to guarantee paperless hiring. According to Jabbar and Abid (2015), the implementation of green recruiting stimulates staff participation in long-term competency monitoring and informs them of company-wide green activities including cutting waste and greenhouse gas emissions. In the end, this makes environmental performance better (Mandip, 2012).

Deshwal (2015) has noted a few eco-friendly hiring procedures, including:

- Online application for resumes
- Utilising employee search engines on business websites.
- Using internet job sites to process onboarding paperwork, including acceptance letters, offer letters, authorisations, and testimonies regarding the experience and educational background of chosen applicants.
- Internet job postings.

According to the study, these procedures guarantee that the least amount of paper is used during the hiring process and are essential for drawing in prospective workers.

7.4.2 GREEN SELECTION

"Green selection process" describes the method of selecting applicants based on their environmental interests and concerns. In an interview, questions have to be relevant to environmental matters. These are some fantastic green selection practises that any organisation can use to select applicants who care about the environment, in addition to the conventional selection criteria related to the specific duties of the post in question (Arulrajah, Opatha & Nawaratne, 2015).

7.4.3 GREEN TRAINING AND DEVELOPMENT

In addition to being extensively used for training in environment management, interactive media and online and web-based training modules should also be extensively used for training in other functional areas (Deshwal, 2015; Rana & Sharma, 2019). Employees can learn how to operate more productively, use resources more sensibly, preserve energy, and have a smaller environmental effect through green training and development (Ullah and Jahan, 2017).

Green training and development practises have been discovered in the study conducted by Arulrajah, Opatha and Nawaratne in 2015. These practises include

- Implementing corporate environmental management programmes to teach and develop personnel with the necessary skills and knowledge.
- Holding organisational seminars and workshops to raise staff members' understanding of environmental issues.
- Offering environmental education with the goal of influencing managers' and non-managerial staff members' attitudes and behaviours.
- Creating a variety of competitive programmes to help staff members and their families understand some important eco-values.

Employees who receive green training and development learn how to work in a way that reduces waste, uses resources wisely, saves energy, and lowers the factors that contribute to environmental degradation. Employees are also given the opportunity to participate in finding solutions for environmental issues.

7.4.4 GREEN PERFORMANCE APPRAISAL

The process of developing "green" behaviour indicators and performance standards, as well as assessing worker performance across the board using "green" targets in the primary performance categories, is known as "green performance appraisal." Green performance management encompasses organisational policies and environmental requirements (Amutha, 2017). Shoeb Ahmad (2015) found that environmental events, environmental responsibilities, environmental policy communication, green information systems, and audits should all be considered when assessing green performance. In addition, he recommended that the performance rating system be updated to incorporate criteria for assessing individuals in areas such as environmental stewardship, diversity, creativity, teamwork, and collaboration. The performance evaluation of organisational managers may be predicated on factors including their ability to raise subordinates' knowledge of environmental issues, motivate them to participate in green initiatives and uphold a positive and supportive work environment.

7.4.5 GREEN COMPENSATION AND REWARD

Since HRM views employees as important assets in the modern era, keeping them on board is crucial for the company. The main factor affecting an employee's retention is a reward and pay management that is appropriate and effective. The company should link its reward programmes with environmental initiatives, such as rewarding recycling and waste reduction efforts, in order to implement GHRM.

Some of the green compensation and reward initiatives have been highlighted by Bangwal and Tiwari (2015) in their research, such as,

- Create packages that recognise the acquisition of green talents.
- The application of financial incentives for environmental management (EM) (premiums, cash, and bonuses)
- Making use of non-cash EM awards (gifts, leaves, and sabbaticals)
- The application of recognition-based EM rewards, such as meals, awards, visibility, outside jobs, and daily praise
- Create negative reinforcements in EM, such as warnings, criticism, and suspensions for infractions.
- Create constructive criticism in EM (feedback)Connect involvement in green projects
- to job advancement (managers progress by helping EM workers).

According to research by Ullah and Jahan (2017), staff accomplishments and green efforts can be recognised with monetary, non-monetary, and recognition-based rewards. Pay raises, cash incentives, and bonuses are examples of monetary rewards; non-monetary benefits could include things like sabbaticals, paid time off, and presents for staff members and their families. Rewards based on recognition could include high management praising staff members in public for their environmental initiatives.

7.5 FACTORS AFFECTING GHRM

Improved working conditions are brought about in the company via GHRM practises. However, it also helps the organisation by improving the financial situation and, above all, the human resources.

In their research article, Vahdati and Vahdati (2018) have discovered a few elements that impact the practises of eco-friendly HRM. These elements are thought to be the most relevant and significant ones for the organisation of today. Since the research by Vahdati and Vahdati is the most recent, just these factors are the subject of this study. The following is a presentation of these factors:

- Personality and upbringing: this category includes things like family, lifestyle, and the way an individual has been raised, a person's character, principles, and worldview.
- Individual knowledge: knowledge, intellect, and degree are components of this category.

- Organisational culture: This element includes the following variables: resource management practises and a culture of protection environment, cultural instruction on eco-friendly conduct, and economical resource use.
- The digital divide: the components of this factor that have been found are: advanced information technology, its impact, properly outfitted workspaces, and its utilisation of information technology, using it responsibly to have more ecologically favourable actions.
- Education and training: The following variables are identified under this factor: Various educational media, centres, and advertisements.

7.6 BENEFITS OF GHRM

This study has identified some of the organisation's non-negligible benefits of GHRM after examining other studies. Researchers have revealed these advantages. These advantages can be summed up as follows:

- GHRM allows employees to engage in environmentally responsible activities, employee morale is raised.
- It is a method of protecting the environment from the damaging effects of organisational procedures.
- It assists organisations in defining environmental issues and in analysing potential remedies.
- It lowers the company's overall expenditures because expenses are mostly determined by the size of the business and the measures made to make it environmentally friendly.
- Organisations can increase their performance with the use of GHRM practises.
- Green business practises create an environment where there is a competitive advantage through environmental and economic sustainability.
- GHRM practises enable the company to secure a competitive edge over its competitors by guaranteeing corporate social responsibility.
- Enforcing regulations like establishing a green corporate space with trees and plants, banning smoking on office property, assuring less paperwork, and offering fresh fruits and vegetables as meeting snacks, among other things, guarantees a healthy work atmosphere.

7.7 CHALLENGES FACED BY ORGANISATION IN IMPLEMENTING GHRM

It is not an easy undertaking to implement GHRM practises in an organisation. The company must overcome a number of obstacles in order to put the practises into effect.

- The company is unable to create a thorough plan for implementing GHRM.
- A lot of organisations find it challenging and complex to use green technologies.

- Green infrastructures for HRM are lacking.
- Measuring the impact of GHRM practises on employee behaviour is challenging.
- Creating a GHRM culture throughout the entire organisation is a laborious and ongoing effort.
- Companies could not be giving their staff enough environmental education and training.
- The organisation can have trouble quickly changing the mindset of its employees from traditional HRM to GHRM.
- Other constraints include a lack of collaboration, a shortage of experts, a lack of acquaintance with the subject among experts, a lack of green culture, and internal literature in the field.

7.8 CONCLUSION

Every organisation needs to maintain a sustainable working environment in order to encourage employment among the general public, making GHRM a key practise. Organisations are more conscious of the need to implement green employment practises these days in order to make the companies more environmentally friendly. The primary emphasis of this study was how the organisations essentially implement green practises for their HRM operations. Additionally, it examined a number of current studies and established the elements, advantages, and difficulties that modern organisations confront while putting GHRM into practise.

REFERENCES

Ahmad, S. (2015). Green human resource management: Policies and practices. Cogent Business & Management. Taylor & Francis Journals, 2(1), 1030817–103, Retrieved from: 10.1080/23311975.2015.1030817.

Ali, M. C., Islam, K. M. A., Chung, S., Zayed, N. M., & Afrin, M. (2020). A study of green human resources management (GHRM) and green creativity for human resources professionals. International Journal of Business and Management Future, 4, 57–67. 10.46281/ijbmf.v4i2.857.

Amutha, V. (2017). A theoretical study on green HRM practices. Global Journal for Research Analysis, 6(11), 2277–8160.

Arulrajah, A. A., Opatha, H. H. D. N. P., & Nawaratne, N. N. J. (2015). Green human resource management practices: A review. Sri Lankan Journal of Human Resource Management, 5(1), 2015.

Bangwal, D., & Tiwari, P. (2015). Green HRM – A way to greening the environment. IOSR Journal of Business and Management (IOSR-JBM) e-ISSN: 2278-487X, p-ISSN: 2319-7668, 17(12), Ver. I (Dec. 2015): 45–53.

Chams, N., & García-Blandón, J. (2019). On the importance of sustainable human resource management for the adoption of sustainable development goals. Resources, Conservation and Recycling, 141, 109–122. 10.1016/j.resconrec.2018.10.006.

Deshwal, P. (2015). Green HRM: An organizational strategy of greening people. International Journal of Applied Research 2015, 1(13), 176–181. [5].

Gill, M. (2012). Green HRM: People management commitment to environmental sustainability. Research Journal of Recent Sciences, 1, 244–252.

Jabbar, M. H., & Abid, M. (2015). A study of green HR practices and its impact on environmental performance: A review. MAGNT Research Report, 3(8), 142–154.

Meily, M., & Saragih, S. (2013). Developing new corporate culture through green human resource practice. The 2013 IBEA, International Conference on Business, Economics, and Accounting. Bangkok, Thailand.

Muisyo, P. K., Qin, S., Ho, T. H., & Julius, M. M. (2021). The effect of green HRM practices on green competitive advantage of manufacturing firms. Journal of Manufacturing Technology Management, 33, 22–40. 10.1108/jmtm-10-2020-0388.

Muster, V., & Schrader, U. (2011). Green work-life balance: A new perspective for green HRM. German Journal of Human Resource Management: Zeitschrift für Personalforschung, 25, 140–156. 10.1177/239700221102500205.

Rana, G., & Sharma, R. (2019). Emerging human resource management practices in Industry 4.0. Strategic HR Review, 18(4), 176–181.

Rana, G., & Arya, V. (2023), Green human resource management and environmental performance: mediating role of green innovation – A study from an emerging country. Foresight. 10.1108/FS-04-2021-0094.

Rana, G., & Sharma, R. (2017). Organizational culture as a moderator of the human capital creation-effectiveness. Global HRM Review, 7(5), 31–37.

Shafaei, A., Nejati, M., & Mohd Yusoff, Y. (2020), Green human resource management: A two-study investigation of antecedents and outcomes. International Journal of Manpower, 41(7), pp. 1041–1060. 10.1108/IJM-08-2019-0406

Ullah, M., & Jahan, M. (2017). Integrating environmental sustainability into human resources management: A comprehensive review on green human resources management (green HRM) practices. Economics and Management, 2(1), 6–22.

Vahdati, S., & Vahdati, S. (2018). Identifying the obstacles to green human resource management practices in Iran. International Journal of Human Capital in Urban Management, 3(1), 9–18.

8 Green Technology and Its Effect on the Modern World

Deeksha Dwivedi and Shivani Agarwal

8.1 INTRODUCTION

The rapid growth of technology has brought about a time of extraordinary advancement but has also resulted in significant environmental issues. Due to this conflict, interest in green technology has increased. This paradigm shift aims to decrease the adverse environmental effects of modern industry (Keller, 2023). This chapter attempts to provide an in-depth analysis of the ground-breaking impacts of green technology on the contemporary world. More importantly, it seeks to explore green technology's conceptual roots and historical progression to establish a fundamental understanding of its significance. It examines how conventional technologies have contributed to environmental deterioration and emphasizes the urgent need for sustainable, alternative measures. This chapter's primary objective is to critically respond to the question: How does the pervasive adoption of Green Technology impact modern life's environmental, social, and economic elements? By examining its scope and limits, the research thus lays the framework for a detailed investigation of this query's influence on the modern world.

The Green technology impacts can be visible with the help of the diagram shown in Figure 8.1

A. ***Environmental:*** The literature on green technology provides a thorough background for comprehending its complex effects on day-to-day life. Green technology, also known as environmental or clean technology, refers to various inventions and methods designed to reduce environmental deterioration and promote sustainability. Its critical importance in solving the worldwide issues of pollution, resource scarcity, and climate change has been highlighted by researchers and academics across the globe (Rana & Arya, 2023; Abdulhayan, 2023). Green Technology is characterized by its dedication to environmental stewardship and symbolizes a paradigm shift away from traditional, resource-intensive methods in favor of more sustainable and regenerative ones.

The environmental movements in the middle of the 20th century are where Green Technology first emerged in history. The oil crisis increased

DOI: 10.1201/9781003458944-8

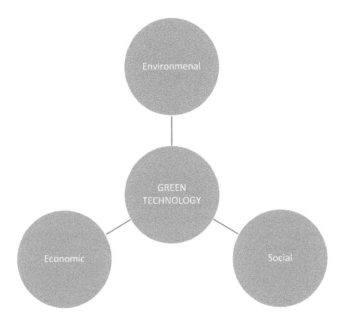

FIGURE 8.1   Impacts of green technology.

interest in alternative, renewable energy sources and a rising understanding of how industrialization has affected the environment. This
sparked the creation of geothermal, wind, and solar technologies, which
have since emerged as pillars of the green technology revolution (Keller,
2023). In addition, developments in waste management, sustainable
transportation, and materials science and engineering have been made
possible by these discoveries.

Critically, the literature emphasizes the need for green technology
adoption due to environmental concerns. Conventional technologies have
taken a tremendous toll on the world's ecosystems since they depend on
fossil fuels and non-renewable resources. The environmental externalities
brought on by these technologies, which range from greenhouse gas
emissions to air and water contamination, have forced us to reconsider
how we approach industrialization (Abdulhayan, 2023). Several solutions
provided by green technology aim to reduce or eliminate these detrimental
effects, protecting biodiversity and natural resources.

B. *Economical:* After Fourth Industrial Revolution, the concept of the "green
economy" came into the picture and was considered as a novel way to
utilize resources. The economic growth of a country was mainly
dependent upon the emergence of new industries, introducing 3 Rs,
zero-emission, green urbanism, etc. Nevertheless, from an economic
perspective, green technologies develop new horizons for growth.

C. *Social:* Green technology has major impacts on society. There are umpteen
perspectives that should be taken care of. First and foremost is "Life cycle
thinking. Life cycle thinking emphasizes on the 'from basic foundation till

grave'". In the technological era, whatever applications and technologies are evolving and emerging. They should be keeping in mind that society should be adaptable to those technologies in a healthy manner. For Example: Chatbot. Another perspective of social impact is the "System approach" which paves the path to develop and use technologies from a holistic approach not as a single or one-time use which further creates trash in the environment (Michael Søgaard Jørgensen & Ulrik Jørgensen, 2009). But it will be helpful to society for a longer duration of time.

8.2 TYPES OF GREEN TECHNOLOGIES

A wide range of cutting-edge approaches known as "green technologies" address environmental problems and advance sustainability. Within this framework, **Renewable Energy** Sources are a crucial category. A notable source of green energy is solar energy, which can be used in photovoltaic cells or concentrated solar power systems. It is a significant participant in the shift to a low-carbon energy environment due to its limitless supply and minimal environmental impact. Similarly, wind energy produced by wind turbines has grown significantly, especially in areas with favorable wind patterns (Keller, 2023). Geothermal energy is another notable aspect of renewable energy technology, produced from Earth's natural heat. Its dependability and negligible adverse effects on the environment make it a competitive alternative to traditional fossil fuels.

Energy Efficiency and Conservation Technologies are essential in reducing resource consumption and environmental effects in addition to renewable energy sources. High-efficiency appliances, insulation for buildings, and smart grid technologies are just a few examples of the inventions that fall under this category. By maximizing energy efficiency, these technologies cut down on waste and greenhouse gas emissions (Abdulhayan, 2023). The pursuit of greener urban landscapes has also made sustainable mobility options essential. Traditional fossil fuel-powered transportation can be replaced with efficient alternatives like electric cars, hybrid systems, and high-tech public transport, which helps to reduce air pollution and dependency on non-renewable resources.

Innovations in waste management and recycling technologies have helped divert waste from landfills and reduce environmental harm. Allowing the extraction of valuable resources from waste streams through advanced recycling processes like mechanical-biological treatment and pyrolysis decreases the demand for virgin materials (Dhinakaran et al., 2020). Additionally, green architecture and buildings have become more prevalent in sustainable urban development. Structures that use fewer resources, produce fewer pollutants, and support a better indoor environment are made possible through passive design techniques, energy-efficient materials, and green infrastructure (Sharma et al., 2021).

Another essential component of green technology **is water conservation and treatment technologies**. These developments, which range from sophisticated waste-water treatment facilities to rainfall harvesting systems, help to manage water resources responsibly (Abdulhayan, 2023). Desalination technologies also offer a sustainable way to increase freshwater supplies for areas struggling with water constraints.

8.3 ECONOMIC AND SOCIAL IMPLICATIONS OF GREEN TECHNOLOGY

Adopting green technologies impacts the economy and society, changing market-places for employment and communities' well-being. In terms of the economy, the shift to sustainable practices and technology has shown its ability to promote expansion and open up new business prospects. For instance, investments in renewable energy projects have produced significant returns, promoting the growth of a robust green economy (Dhinakaran et al., 2020). Additionally, adopting energy-efficient practices and technology in industries like manufacturing and construction has reduced costs for businesses and homes (Sharma et al., 2021). In addition to reducing environmental hazards, this transition to green technology encourages innovation, boosting economic resilience and competitiveness on a global level.

The incorporation of green technologies has significant social repercussions in addition to economic advantages. The influence on job development and worker transformation is one of the most noticeable. Particularly in the field of renewable energy, job prospects are abundant and span a wide range of positions, from technicians to engineers to project managers. In addition to meeting job needs, this encourages skill diversification and the expansion of a long-term workforce (Abdulhayan, 2023; Rana & Sharma, 2021). Additionally, the creation and upkeep of green infrastructure, such as energy-efficient buildings and public transit networks, helps to revitalize towns by improving social cohesiveness and quality of life.

Furthermore, improving social equity and environmental justice depends heavily on green technologies. Historically, environmental pollution and deterioration have fallen disproportionately on underprivileged groups. Green technologies can redress this imbalance by lowering pollutants, enhancing air quality, and supplying cleaner, more affordable energy sources (Keller, 2023). Underserved communities will benefit from a sustainable future thanks to initiatives focusing on fair access to green technologies. Reduced exposure to dangerous contaminants and a more sustainable, habitable urban environment also benefit populations' health and well-being.

8.4 ENVIRONMENTAL IMPACT ASSESSMENT OF GREEN TECHNOLOGIES

A viable strategy for reducing environmental damage and maintaining natural ecosystems is the adoption of green technologies. Conducting a thorough Environmental Impact Assessment (EIA) to quantify the consequences of these technologies on a quantitative and qualitative level is a vital aspect of this transition. The significant decrease in greenhouse gas emissions that renewable energy sources like solar and wind experience compared to traditional fossil fuels is an essential advantage of green technologies (Dhinakaran et al., 2020). According to studies, the widespread use of these technologies has significantly reduced emissions of carbon dioxide and other greenhouse gases, aiding in the fight against climate change on a global scale.

Green technologies have a favorable effect on biodiversity and resource conservation in addition to energy production. For instance, by diverting items from landfills and lowering the need for virgin resources, waste management and recycling technology improvements have considerably lessened the strain on natural resources (Nalley, 2021). Additionally, environmentally friendly mobility choices, such as electric cars and effective mass transit, help minimize emissions and traffic, improving urban air quality and lowering noise pollution. Throughout a structure's life, green building techniques, defined by energy-efficient designs and materials, help reduce energy consumption and environmental impact.

8.5   POLICY AND REGULATORY FRAMEWORK FOR ADOPTION OF GREEN TECHNOLOGIES

A strong and encouraging policy and regulatory framework is essential for the broad adoption of green technologies. Governments have a crucial role in creating a favorable climate for developing, adopting, and expanding these technologies. The creation of incentives and subsidies to promote the adoption of green technologies is one of the most critical techniques used (Nalley, 2021). For the purpose of offsetting the upfront costs of implementing renewable energy sources, tax credits, feed-in tariffs, and subsidies have been crucial in encouraging people, organizations, and communities to invest in sustainable practices.

Governments also use various regulatory measures to compel and promote green technologies. This entails establishing standards for energy efficiency across diverse industries and targets for renewable energy and emission reduction. These goals show a commitment to shifting to a more sustainable energy environment and serving as clear benchmarks for success (Dhinakaran et al., 2020). Additionally, supporting sustainable building and development depends on building rules and standards prioritizing energy efficiency and green building techniques (Sharma et al., 2019).

International accords and treaties also shape the policy environment for green technologies. The Paris Agreement provides a global foundation for cooperation in addressing climate change, encouraging countries to set high goals for decreasing emissions and boosting renewable energy sources (Nalley, 2021). These international agreements speed up the global shift to green technologies by facilitating knowledge sharing, technological transfer, and cooperative research projects.

8.6   FUTURE TRENDS AND INNOVATIONS IN GREEN TECHNOLOGY

With several new trends and developments on the horizon, the world of green technology is set for a dynamic evolution. The ongoing development of renewable energy technologies is one of the most noticeable trajectories. Innovations in energy storage, such as grid-scale storage options and next-generation batteries, are positioned to overcome the problems with solar and wind energy's cyclical nature (Nalley, 2021). Furthermore, advances in nanotechnology and sophisticated materials have the potential to boost the effectiveness and scalability of renewable energy systems. Emerging technologies, such as ocean energy and biofuels derived from algae, give intriguing new opportunities for extending the portfolio of renewable energy sources beyond conventional ones.

Often referred to as **"smart green technology"** the fusion of information technology and environmentally friendly practices is another significant development. Real-time monitoring and resource utilization optimization are made possible by combining sensors, data analytics, and artificial intelligence. For instance, intelligent grid technologies enable dynamic energy distribution, enabling more effective and flexible management of electricity resources (Dhinakaran et al., 2020). The growth of Internet of Things (IoT) devices and intelligent building technology makes a more sustainable and comfortable living and working environment possible (Prakash et al., 2021).

The electrification of various transit systems is expected to transform mobility in transportation. Thanks to improvements in battery technology and charging infrastructure, electric vehicles (EVs) are expected to become more widely available and reasonably priced (Singh & Srivastava, 2022). Additionally, integrating autonomous and shared mobility solutions can improve transportation systems even further by lowering traffic, pollution, and environmental effects in general.

Biotechnology and biomimicry are proving to be practical innovations in green technology. To develop more sustainable materials, processes, and systems, bio-inspired design principles are being used. These principles are derived from nature's responses to complex problems (Nalley, 2021). These technological advancements, which range from bio-based polymers to biomimetic architecture designs, have the potential to lessen the environmental effect and resource usage drastically.

8.7 CONCLUSION AND RECOMMENDATIONS

In conclusion, this study has thoroughly examined Green Technology and its significant influence on contemporary society. The analysis of numerous green technologies, including renewable energy sources, energy efficiency measures, waste management techniques, and environmentally friendly modes of transportation, highlights how crucial these technologies are in preventing environmental deterioration and advancing sustainability. The environmental impact assessment identified benefits that may be seen and measured, such as decreased greenhouse gas emissions, better air quality, and resource conservation (Singh & Srivastava, 2022). The consequences for the economy and society also showed the potential for increased community well-being, job development, and economic progress. The policy and regulatory framework section highlighted the importance of government programs, rewards, and international agreements in creating an environment where green technology adoption is possible.

With new developments in renewable energy, innovative technology integration, and biologically inspired solutions poised to transform businesses and urban environments, the future of green technology is promising. It is essential to keep funding research and development and promote partnerships between the public, private, and non-profit sectors to realize these potentials fully. Policies should also prioritize financial incentives for using green technologies and establish challenging goals for deploying renewable energy sources and reducing emissions (Dhinakaran et al., 2020). Campaigns for public education and awareness are also crucial for promoting a culture of sustainability and boosting consumer demand for green technologies.

REFERENCES

Abdulhayan, S. (2023). Green blockchain technology for Sustainable Smart Cities. *Green Blockchain Technology for Sustainable Smart Cities*, 237–262. 10.1016/b978-0-323-95407-5.00014-1

Dhinakaran, V., Surendran, R., Varsha Shree, M., & Gupta, P. (2020). Study on electric vehicle (EV) and its developments based on batteries, drive systems, and charging methodologies in Modern World. *Electric Vehicles*, 103–118. 10.1007/978-981-15-9251-5_6

Jørgensen, M. S. & Jørgensen, U. (2009). Green technology foresight of high technology: a social shaping of technology approach to the analysis of hopes and hypes. *Technology Analysis & Strategic Management*, 21(3), 363–379. 10.1080/09537320902750764

Keller, J. (2023). 2023 IEEE Green Technology Conference sponsors. 2023 IEEE Green Technologies Conference (GreenTech). 10.1109/greentech56823.2023.10173848

Nalley, E. A. (2021). Technology supporting Green Chemistry in chemical education. *Green Chemistry and Technology*, 27–44. 10.1515/9783110669985-003.

Prakash, C., Saini, R., & Sharma, R. (2021). Role of Internet of Things (IoT) in Sustaining Disruptive Businesses. In R. Sharma, R. Saini, C. Prakash, & V. Prashad, *Internet of Things and Businesses in a Disruptive Economy* (1st ed.). Nova Science Publishers.

Rana, G. & Arya, V. (2023). Green human resource management and environmental performance: mediating role of green innovation – a study from an emerging country. *Foresight*, Vol. ahead-of-print No. ahead-of-print. 10.1108/FS-04-2021-0094.

Rana, G. & Sharma, R. (2021). Employer branding: attracting and retaining employees for sustainable development in disruptive economy. *World Review of Science, Technology and Sustainable Development*, 17(4), 319. 10.1504/wrstsd.2021.117839.

Sharma, R., Saini, R., Prakash, C., & Prasad, V. (2021). *Internet of Things and Businesses in a Disruptive Economy*, 1st edn. Nova Science Publishers.

Sharma, R., Singh, S., & Rana, G. (2019). Employer Branding Analytics and Retention Strategies for Sustainable Growth of Organizations. In *Understanding the Role of Business Analytics*, pp. 189–205. Springer. 10.1007/978-981-13-1334-9_10.

Singh, A. & Srivastava, Y. K. (2022). Green innovation and green technology. *Patent Law, Green Technology and Innovation*, 14–59. 10.4324/9781003319467-2

9 Green Bonds

Accelerating Green Finance Towards Sustainable Economic Development

Varsha Gupta and Abhishek Kumar

9.1 INTRODUCTION

The growth story of India has been impressive, and increased employment opportunities over the years have permitted millions of people to elevate themselves above the poverty line (WorldBank, 2014). Since 2000, India's GDP has grown on average at 7%, with some years exceeding 9%. The Indian economy reported an average growth rate of 7% between 2008 and 2011 (World Economic Statistics, 2015). A growing Gross Domestic Product (GDP) is crucial to India's ambition of becoming a five trillion-dollar economy by 2024. Climate change is perhaps the greatest threat to this (Jain, 2020).

As a result of rapid industrialization and economic growth, unhealthy air and water have become a problem which in turn has resulted in negative effects on human well-being, a high infant mortality rate, and many more issues (Chandra, 2015). Throughout the journey of industrialization and modernization, the country has struggled with environmental issues such as pollution, deforestation, and global warming, all of which have adversely impacted its economic and health development.

In 2023, India experienced its hottest February since 1901, the first year the Meteorological Department started keeping weather records. The regular occurrence of acute changes in weather events is likely to worsen their impacts on human well-being. As stated in the World Bank report (2023) on development and a changing climate, India is the world's most populated country having around 1.4 billion people, and thus its carbon intensity directly impacts global emissions. India released 3.9 billion CO_2 equivalent tonnes of greenhouse gases in 2021 and accounted for being world's third-largest emitter of CO_2. In comparison with a world average of 6.9 CO_2 equivalent tonne and 17.5 CO_2 equivalent tonne in the United States, India's greenhouse gas emissions per capita were around 2.8 CO_2 equivalent tonne.

According to the Department of Economic and Policy Research (DEPR), rising temperatures and altered monsoon rainfall patterns are projected to cost the Indian economy 2.8% of its GDP by 2050 and lower living standards for nearly half of the

DOI: 10.1201/9781003458944-9

population. Without adequate mitigation policies, India may lose between 3% and 10% of its GDP annually by 2100 as a result of climate change (Sultana, 2023). As a result of climate-related events in 2019, India lost nearly \$69 billion, compared with \$79.5 billion lost over the years 1998–2017. Over 1.8 million people were displaced and 1,800 people died as a result of floods that hit nearly 14 states in India in 2019. A total of around 12 million people were adversely affected by the heavy rains in 2019 during the monsoon season, resulting in economic losses estimated at approximately \$10 billion. Considering the severity of climate change's impact on the economy, The International Labour Organization (ILO) has projected a loss of 34 million full-time jobs by 2030 in the farming community especially, if these issues are not addressed properly (Bansal, 2022).

The Reserve Bank of India (RBI) and other financial sector regulators have gradually focused their attention on assessing the potential threats to financial stability that result from the effects of climate change through a variety of policy strategies, such as promoting green financing, starting the issuance of green bonds, and creating a platform for trading carbon credits. India still has a long way to go even though it has started the process of green finance and has been ahead of other nations. In order to attract additional capital, the country needs a successful integration of a competitive tax structure, reliable financial products, and unambiguous environmental standards. Effective collaboration between decision-makers, private, and public organizations, as well as a constant focus on the broader picture, can help in attaining the goal of Sustainable economic growth and green funding (Bansal, 2022).

9.2   EMERGENCE OF NEED FOR GREEN FINANCING AROUND THE WORLD

Variations in climate globally are largely triggered by the emission of carbon dioxide. In order to prevent the worst impacts of climate change, emissions must urgently be reduced. All countries, individuals, and governments should shoulder this responsibility collectively.

Until the mid-20th century, fossil fuel emissions were relatively low and even grew slowly until the Industrial Revolution. With time, emissions from fossil fuels have increased, and emissions from land use change have shown a downtrend since 1946 when land use change was recorded at 5.17 in comparison to 4.64 emissions from fossil fuels. Later the growth in emissions became a matter of concern and 6 billion tonnes of CO_2 emitted in 1950 had almost quadrupled, reaching more than 22 billion tonnes in 1990s. The emission has crossed the level of 45 billion in the year 2016 from fossil fuel only (Ritchie & Roser, 2020)

As shown in Figure 9.1, total CO_2 emissions increased from 30.57 billion tonnes to 50.05 billion tonnes just in the span of 16 years, although the growth of emissions has slowed over the last 5 years, they are still at a high level. Fossil fuel has been majorly responsible for CO_2 emissions around the world. Up until the end of the 20th century, Europe and the United States were majorly responsible for global emissions. As a matter of fact, they accounted for more than 85% of emissions till the end of 1950s, but gradually the scenario changed, and currently they account for around one-third of CO_2 emissions. As shown in Figure 9.1, 218 countries around

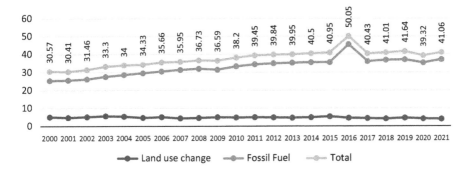

FIGURE 9.1 Status of carbon emissions around the world (Data in Billion Tonnes).

Source: Author Made, based on data retrieved from https://ourworldindata.org

the world are responsible for 37.12 billion of carbon emissions from fossil fuels in the year 2021, which has increased significantly from 25.45 billion tonnes in the year 2000, indicating an absolute change of 45.8%. The United States, China, Russia, Japan, and India were the five leading countries responsible for carbon emissions from fossil fuels in the year 2000.

Carbon emissions by the United States decreased from 6.02 billion tonnes in the year 2000 to 5.01 billion tonnes in the year 2021, while that of China (2.15x) and India (1.76x) has upsurges considerably as shown in Figure 9.2. Currently, China alone accounts for 31% of the entire carbon emission from fossil fuels around the world.

As per the report published on ourworldindata.org, greenhouse gas emissions were 49.4 billion tonnes of CO_2 in the year 2016 out of which the energy sector alone accounted for 73.2% of greenhouse gas emissions followed by 18.4% share of agriculture, forestry, and land use. Further, the industry sector is liable for 24.2% of the consumption of energy, transportation for 16%, and infrastructure (building) for 17.5%, which indicates that the industrial sector is majorly accountable for greenhouse emissions.

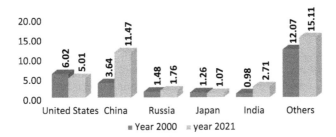

FIGURE 9.2 Country-wise CO_2 emissions from fossil fuels (Data in billion tonnes).

Source: Author Made, based on data retrieved from https://ourworldindata.org

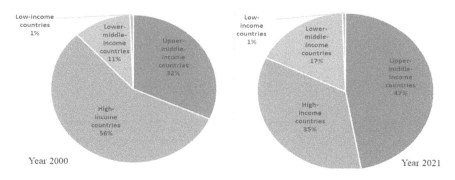

FIGURE 9.3   Income category-wise changeover in shares of carbon emission.

Source: Author's own.

Further, as shown in Figure 9.3 high-income countries, which were accountable for 56% of carbon emissions, have reduced their stake to 35% indicating adoption of sustainable practices. Just opposite to this, the increased pace of industrialization has accounted for an increased share in carbon emissions by upper-middle-income countries and lower-middle-income countries in the 21st century.

Osadume and University (2021) in their study conducted in West Africa between 1980 and 2019 highlighted a positive correlation of economic growth with carbon emission. The study further proposed to promote energy-efficient projects and sustainable development and to avoid the negative impacts of carbon emissions on human well-being and the growth of the country. For promoting sustainable development, the contribution of the financial sector is imperative as the sector facilitates the introduction and promotion of innovative tools and techniques towards the evolution of a low-carbon economy.

By collecting and transferring capital toward renewable energy, electrification, and improved operational efficiency, public and private concerns can help promote sustainability around the globe.

9.3   LITERATURE REVIEW

Dubash and Rajan (2001) stated that green, ecological change is becoming increasingly important to business organizations, while on the other hand, business entities are being forced to issue customized financial instruments, making green bonds an important concept in emerging economies.

Kim (2015) reported three major benefits of issuing green bonds over traditional bonds. The promptness of investors to invest in eco-friendly projects and fight against the adverse effects of climate change serves as one of the main factors for the growth of the green bond market. Furthermore, the stated bonds serve as a fascinating marketing tool for all stakeholders because of their environmentally friendly nature and cost-efficient issuance of green bonds compared with traditional bonds, which are other two benefits associated with green bonds markets

Peri M. (2019) highlighted the quantum of money required for financing the projects supporting a carbon-efficient economy so as to achieve the stated goals of the Paris Agreement. Further, there has been a significant increase in the popularity of green and sustainable bonds, as a means of financing environmentally friendly and sustainable development projects. In spite of the growth in the green bond market, their relevance over and above other categories of projects without having "greenness" is a matter of concern. The study covered 121 European green bonds issued between 2013–17 and propounded the significance of green bonds in terms of cost-efficient purchase compared with other traditional bonds.

Sartzetakis E. (2020) highlighted the potential role of green bonds in financing the projects and transforming the country into a low-carbon economy. The research stressed the significance of the central bank in addressing the issues related to environmental issues and suggested the usage of green bonds for financing long-term infrastructure projects to control carbon emissions in the economy. The study has summarized the growth of the green bonds market over the last year and considered obstacles and challenges in financing as a boost to the bond issuance market.

Stojanović (2020) examines green bonds as an emerging and potential source of financing renewable energy projects. Over the past few years, the market for green bonds has grown significantly due to greater investor awareness of climate change. A financial market's guidelines and standards are essential to enhancing the market's development and achieving green finance goals. The study takes a theoretical approach to the green bond market in order to identify the main obstacles that prevent many nations from utilizing this expanding form of funding for renewable energy. The study highlighted a non-suitable institutional structure for managing the green bond market, the requisite minimum volume of issuance, and high transaction cost as the three major factors hindering the growth of the green bond market. The study recommended the implementation of correct policies for promoting green bonds to finance climate change projects and mitigation projects.

Mecu et al. (2021) reported that over the past decade, around 12% to 42% of all investments made by investors have been placed in green, social responsibility, and sustainable assets. The study further stated that the socioeconomic injustices and the threats associated with climate change endanger society. Green bonds and social bonds are serving as important tools to enable financing the environmental and social transformation processes and projects.

A study indicating the positive impact of the issuance of green bonds on the stock market propounded that investors consider companies as better investment avenues that indulge in raising funds from green bonds and investing in environmentally friendly projects (Baulkaran, 2019).

The above-cited literature suggests that the green bond market is becoming popular and the issuance of green bonds and green finance has become the need of the time despite the fact that the green bond market is at its infancy stage. A few pieces of literature have claimed the benefits of green bond issuance in terms of increased stock-price valuation of companies raising funds for reducing carbon emission in the country, while others highlighted the challenges and high cost of financing serving as an obstacle in the growth of this market. Moreover, very few studies indicating the current status of the green bonds market in India are available, so the current study focused on analyzing the current scenario of the green bond market and challenges encountered by issuers in India.

9.4   OBJECTIVES OF THE STUDY

To analyze the current scenario of the green bond market, as one of the instruments for financing energy-efficient and sustainable projects.

To explore the challenges encountered by issuers in harnessing green bonds.

9.5   METHODOLOGY

The study is basically exploratory in nature, and the literature based on previously conducted studies and articles/reports published on websites has been considered for relevant information about climate issues arising around the world and India and the impact of climate change and carbon emission on economic development over the years. Further, data and relevant information regarding the emergence of green financing with a special focus on green bonds, serving as an important tool for accelerating environment-friendly projects and the growth of the economy, their related benefits and challenges have been collected from World Bank reports, RBI bulletins, Articles of newspapers, and other informative websites. The study basically covers the time period related to green bond issuance and expansion from year 2008 to 2021.

9.6   GREEN FINANCE IN INDIAN CONTEXT

As part of transitioning to a low-carbon or green economy, substantial amounts of fresh capital investment are needed, in particular sustainable financing, which supports and promotes environmentally friendly activities and innovations in fields of renewable energy, clean transportation, and conservation. A green finance program is a program that provides financial support for projects, businesses, and initiatives that have a constructive effect on the environment and humanity. Green finance refers to loans or investments that promote environmentally friendly activities. For instance, green finance might be used to buy products and services that are eco-friendly or to build green infrastructure. Fostering the usage of environmentally friendly goods and services and transitioning to a low-carbon society will result in great green multiplier effects, creating a win-win platform for both the economy and the environment. The purpose of green investing is to reduce environmental risks, improve ecological integrity, and improve human well-being while conserving the environment and the natural capital at the same time. Over the past few decades, green financing has grown significantly in India, starting in the early 2000s.

Though there is no specific definition of green finance that has international agreement, still green financing is referred to as financing environment-friendly technologies, projects, businesses, and ideas. In general terms, green finance indicates all financial products and services like credit cards, loans, insurance, and bonds, which are focused on promoting a clean and carbon-free environment.

As defined by the UN Environment Program, "Green Financing" refers to the inflow of funds from the private and non-profit sectors to support sustainable development.

Sustainable development and environmental awareness began to grow in India in the early 2000s. Companies began integrating environmental considerations into their operations as the concept of Corporate Social Responsibility (CSR) gained attention. In spite of this, there were few green financing mechanisms. There was a significant push to promote renewable energy in India during the 2010s. Solar energy development became a national priority with the launch of the National Solar Mission in 2010. Green financing grew as a result of this initiative, which attracted domestic and international investments in solar projects. As a result, financial institutions and banks in India commenced practices of sustainable banking by integrating socio-environmental factors into their lending and investment activities and developed green loan products, including loans for renewable energy projects and energy-saving technologies. A few of the categories of green finance are as follows.

9.7   TYPES OF GREEN FINANCE

9.7.1   GREEN MORTGAGES

It stands for better financing terms for home buyers when the underlined property has a huge environmental sustainability score or the purchaser himself agrees to improve the environmental performance of the property.

9.7.2   GREEN LOANS

It is a loan offered to support solar panels installed by households, electric automobiles, projects indicating energy efficacy, and many more environmental initiatives.

9.7.3   GREEN VENTURE CAPITAL

Companies that develop sustainable technologies and clean energy solutions are often backed by venture capital firms, which invest in startups and early-stage companies through equity financing.

9.7.4   GREEN BANKING

An innovative and mission-driven institution, the Green Bank aims to speed up the changeover to clean energy and fight climate change using advanced and improved financing. In contrast to traditional banks, green banks are focused on deploying clean energy rather than maximizing profits. Their goal is to engage in market research and develop a pipeline of clean projects. Almost all Green Banks aim to reduce climate change, although many also serve low-income communities or improve resiliency. Green banks use financing instead of grants. The purpose of financing is to ensure that capital will eventually be repaid, maximizing the impact of each dollar that a green bank deploys. Therefore, green banks focus on markets with a high potential for return.

9.7.5   Green Bonds

Green bonds are fixed-income security issued by government agencies, institutions, or corporations to raise capital for eco-friendly projects. Green bonds are financial instruments offering static returns meant to raise finance for projects related to renewable energy, energy competence, sustainable agriculture, and other eco-friendly initiatives. They typically come with tax enticements to augment their appeal to investors.

A green bond is issued by a sovereign entity, an intergovernmental group or alliance, or a corporation in order to fund projects that have an optimistic impression on the environment. Though, the World Bank issued the first official Green Bond in 2008 and during that year, around $ 1.57 billion worth of funds were being raised through the issuance of Green Bonds. The World Bank's green bond helped to increase public awareness of the problems caused by climate change while also showing how stakeholders may finance climate solutions through secure investments without sacrificing financial gains. In conjunction with ICMA, the International Capital Markets Association, it formed the basis for the Green Bond Principles. In addition to stressing the value that bonds could create, it stressed the necessity for a more transparent bond market. A commitment of $1.2 billion and a disbursement of $0.8 billion were made in FY22. 126 projects have been funded in 35 countries since 2008, committing $19.5 billion and disbursing $12.3 billion (Green Bond- World Bank, 2022). As stated in the report, the majority of funds were committed and disbursed in the renewable energy and energy efficiency sector. Around 35% in renewable energy and energy efficiency sector, followed by 29% in clean transportation and 12% in agriculture, land use and forestry sectors have been disbursed. Further, considering regional disbursement, East Asia and the Pacific followed by Latin America and the Caribbean have been leading ahead.

A study by the City UK and BNP Paribas (March 2022) shows that global green financing has grown over 100x in the span of the last ten years. The report stated that investing in green bonds and loans, as well as equity funding through initial public offers aiming at green projects, soared to US$ 540.6 billion in the year 2021 from US$ 5.2 billion in the year 2012. It demonstrates how governments and corporations are trying to curb carbon emissions and reach climate goals in response to the rise in issuance. In the period between 2012 and 2021, green bonds accounted for 93.1% of all green finance worldwide. A total of $511.5 billion in green bonds were issued in 2021, compared to $2.3 billion in 2012. China followed by the US, France, and Germany accounted for 13.6%, 11.6%, and 10% in each issuance of green bonds between the stated time. In both the public and private sectors, green bond issuance has grown steadily. Despite having a dominant role for public issuers in 2012, corporate issuers have now dominated green debt capital markets since 2015. In the past ten years, nine of the ten largest green bonds have been issued by European countries, of which the UK ($13.7bn) and Italy ($10.1bn) issued the largest. In the period between 2012 and 21, four of the ten largest corporate green bonds were issued by US companies, the largest issuance being Apple's ($12bn, in seven tranches) in 2016 (Green finance: A quantitative assessment of market trends, 2022).

9.8 GREEN BONDS TO SUPPORT THE CLEAN ENERGY DEVELOPMENT IN INDIA

Financing and supporting the renewable energy project had been a major matter of concern for the government considering the relevance and impact on the environment. The green bond market can serve as a source of finance, as per the report of the Council on Energy, Environment and Water:

1. **Green bonds facilitate in promoting funds and investor base:** Green bonds are specifically designated for initiatives that improve the environment, such as renewable energy projects. These bonds appeal to investors because they can see how their money is being utilized to fund environmentally friendly projects. To meet India's goals for clean energy, a variety of tactics and methods are needed to raise sufficient money in a timely manner. In India, finance for infrastructure has typically been provided by organizations like banks, NBFCs, and financial institutions. Traditional funding sources like domestic bank loans are insufficient to support capacity increase due to the significant expenditure needed to produce renewable energy. A wider investor base, including pension funds, insurance companies, etc is needed to benefit from new creative financial products, such as green bonds, in order to meet India's climate and renewable energy commitments.

   Additionally, green bonds can draw international investment into India's renewable energy industry. International investors are frequently keen to invest in initiatives that support sustainable development goals, particularly foreign institutional investors and development finance institutions.

2. **Provide access to long-term and cost-efficient capital:** Compared to other bonds and financial institutions like commercial banks, green bonds offer funds at a cheaper rate. Due to the lengthier payback period typically associated with renewable energy projects, banks have difficulty participating in long-term projects. Since the interest rates in India are high, most green initiatives require a lot of upfront capital investment. Bonds with a low-risk-return profile make green energy projects more affordable and reliable regardless of governmental policy. By providing scalable low-cost capital through institutional investors, green bonds can facilitate global financing for renewable energy projects. Any minor cost increases in accounting and certification requirements will be quickly reduced through diversification, increased demand, and increased green bond market development.

3. **Liquidity and Refinancing:** In order to refinance lending institutions, green bonds are used, and the money raised can then be used to fund renewable energy and other sustainable development initiatives. In order to fulfill all nations' climate commitments under the 2015 UN climate agreement in Paris, an estimated $1 trillion annually is needed. Total

86% of global financial flows are anticipated to come from private sector investments. In order to achieve low-carbon development goals, significant private funding would be needed. Green bonds, for instance, are appealing financing methods that can be used more widely since they can access private sources of finance and have extra benefits.

4. **Facilitate gratification of climate obligations:** Green bonds can help the international investor community fulfill the rising demand for climate-friendly investments. Investors are becoming increasingly concerned about incorporating ESG considerations into their portfolio of investment and an approximately publicly disclosed amount of $45 trillion has already been committed to such projects through green initiatives, and further green bonds can serve as an important financial tool in leveraging the market

9.9   GROWTH OF GREEN BOND MARKET IN INDIA

The significance of green finance was felt long back in 2007 in India. Reserve Bank of India, through its notification RBI/2007–08/216, emphasized paying attention to climate changes resulting from global warming with reference to sustainable development (Corporate Social Responsibility, Sustainable Development, 2007). In 2008, the government outlined the policy outline of the Action Plan on Climate Change to mitigate the impact of climate variation on the economy. The Climate Change Finance Unit (CCFU) was incorporated in the year 2011 with a vision to synchronize the working of different financial institutions involved in operations related to green finance in India. Implementation of the sustainability disclosure requirements proved to be a major strategic move since 2012 as part of these initiatives. Since 2012, the regulatory body of the stock market, the Security and Exchange Board of India (SEBI) has made it obligatory for the top hundred companies based on their market capitalization, listed at BSE and NSE to prepare and submit business responsibility reports annually. SEBI published guidelines in May 2017 specifying disclosure requirements for green bonds. moreover, as part of the Companies Act of 2013, the Ministry of Corporate Affairs requires companies to report on their Corporate Social Responsibilities (CSR). India has taken a number of policy initiatives to encounter and fulfill its commitments under the Paris Agreement of reducing greenhouse gas emissions by 33–35% below the level in 2005 and procurement of 40% of its electric power capacity from non-fossil sources by 2030 (RBI Bulletin, 2021). There has been a significant contribution of financial sectors towards financing various projects supporting the production of electric power from non-fossil sources.

Although issuance of bonds for financing projects has started long back. The concept of green bonds emerged in 2014 in India. Aligned with the green bond concept, certain bonds were issued by Indian corporations and governments from January 21, 2015, irrespective of whether they were green bonds. The first bank in India to issue a green bond was YES Bank in the year 2015. This was an important phase in India's transition to green funding. The success of Yes Bank in terms of issuance of green bonds, "debt products created especially to finance ecologically friendly projects" encouraged other financial institutions to foster similar funding strategies. Afterward with implementation of guidelines for the issuing and listing of green bonds in India in 2017 by SEBI, offered a uniform framework for corporate and institutional issuers, ensuring responsibility and transparency in the use of revenues. With time, the Reserve Bank of India (RBI) and SEBI kept on refining the policy guidelines including the introduction of new reporting guidelines for environmental, social, and governance (ESG) parameters. In spite of various steps taken by regulatory bodies, a report of the Climate Policy Initiative, which analyzes and advises on green investments, has been released indicating India's first-ever attempt to track green investment flows. According to the organization, green finance in India was only INR 309,000 crore (nearly $44 billion) during the year 2019–2020, less than a fourth of the country's needs. Furthermore, India required approximately 164.25 lakh crores ($2.5 trillion) between 2015 and 2030 and 11 lakh crores ($170 billion) per year during the stated time (Khanna, 2022).

As of February 12, 2020, India had floated green bonds totaling around US$ 8 billion since January 1, 2018, which is approximately 0.7% of all the bonds that have been issued on the Indian financial market. Compared with several advanced and emerging economies, India maintained a favorable position despite issuing a very small amount of green bonds since 2018 (RBI Bulletin, 2021). India stood at 7th rank by issuing 22 bonds of $7,792 million through ESG (Environmental, Social and Governance) and green bonds.

Although, around $43 billion apart from Sovereign green bonds have been raised by corporate entities from January 2014 to March 2023, towards fostering investment into renewable energy projects in the country. The money has been raised both from labeled and unlabeled green bonds. "Unlabeled" bonds are those issued by issuers that operate exclusively in the renewable energy sector, but are not certified as "green". Within the stated time, nearly 80% of the green bonds in India were issued by domestic power producers, wherein Greenko, ReNew, NHPC, Adani Green, and Continuum being among the top five issuers of these bonds. The share of these top five players has reduced to 62% over the last three years (BloombergNEF report, 2023).

The majority of green bonds floated in India have a maturity period of between five to ten years and are denominated in US$. As stated in RBI bulletin 2021, Green bonds denominated in Indian currency with long-term maturity (more than ten years) offered higher coupon rates than traditional bonds (Both Govt and Privately issued bonds), as against the bonds denominated in US$. So, bonds denominated in US dollars are termed to be more cost-efficient for the issuer than those issued in domestic currency. In spite of the high coupon rate offered to investors, the stock price of such listed bond projects/companies was tradable at a premium value in the stock market (Baulkaran, 2019).

Green bonds denominated in Indian Currency amounting to INR 75.5 billion have been floated by seven issuers since 2015, wherein National Thermal Power Corporation (NTPC) and Indian Renewable Energy Development Agency Ltd (IREDA) collectively account for the issuance of more than 50% (INR 46.5 billion). Further bonds denominated in US dollars amounting US$ 1,750 million were issued by Greenko (2016, 2019), followed by US$ 1600, US$ 862, US$ 850, US$ 750 by ReNew Power (2017, 2019, 2020), Adani Green Energy (2019), Azure Power

(2017, 2019) and State Bank of India (2019, 2020) respectively have been issued in India since 2015 (Bagaria, 2020).

Financing India's green projects will still require considerably more funding, and issuers have met several issues in leveraging green bonds. Despite the emergence of green bonds, investors and issuers faced several challenges. Some of the major challenges in the issuance of green bonds in India are as follows:

1. **Lack of awareness and Compliance concerns:** Various participants in the green bond market often misunderstand green bonds, their benefits, and principles, which is one of the major obstacles to the green bond issue in India. Potential issuers may be discouraged due to compliance concerns and a lack of investor demand. It is crucial to educate investors on the purpose, structure, and benefits of green bonds in order to encourage more investment in this market.

2. **Regulatory Environment:** Green bonds are an imperative tool to finance green projects. A supportive regulatory environment helps to ensure that green bonds are issued responsibly and used for their intended purpose. This in turn helps to attract investors, which is essential for the success of green bonds. The lack of such a regulatory environment serves as a threat for issuers of green bonds. Although SEBI has taken certain steps by introducing certain guidelines in this direction, still to foster investors' confidence, more refinement and alignment are needed with international standards such as the Green Bond Principles and Climate Bonds Initiative.

3. **Shortage of eligible green projects** also serves as one of the major obstacles to the growth of green bonds. A number of sustainable projects, such as renewable energy and energy efficiency, are planned for the country, but the majority of these will face challenges related to development, finance, and bankability. An investor cannot evaluate the potential of a new project, its effect on the environment, and the long-term sustainability of the same due to its uncertain nature and lack of standardization. Project developers and government agencies can be encouraged to prioritize green initiatives, expedite the project approval process, and offer technical assistance and financial support in order to increase qualified project supplies.

4. **Pricing of the issue** is another problem as evaluation by external parties, certifications, and reporting adds towards the cost of the issue, which in turn discourages issuers. Various financing methods are needed to meet this challenge, such as green bond insurance, green bond guarantee programs, concessional funding, tax benefits, or carbon credits. These techniques can reduce expenses making green bonds more profitable for issuers.

5. **Investors' appetite and market liquidity:** It is highly desirable for green bonds. Despite the high rate of interest from institutional and retail investors, the Indian market lacks a significant number of investors seeking to invest money. Investors are generally more concerned about risk-return profile, transparency, innovations, and longer tenure of green bonds while comparing them with traditional fixed-income securities. A bigger pool of

investors is ensured by the adoption of consistent reporting and independent verification, which can boost investor confidence and market liquidity.

6. **Market fragmentations:** Due to different countries and organizations adopting their own guidelines and standards, there may be fragmentation in the green bond market. Because of this, investors and issuers cannot compare opportunities easily.

In spite of various challenges, the green bond market is gaining popularity year on year, though, the practices of greenwashing also have mounted in the market simultaneously. Falsifying, misleading, unsubstantiated, or otherwise incomplete statements about a product's sustainability are known as "greenwashing". So, it is a practice of misleading or exaggerating the ecological benefits of a product & service, or company, in order to portray an image of more environmental responsibility owned by the company raising funds through green bonds. In the green bond market, greenwashing has negative consequences that include:

1. **Loss of investors' trust:** A green bond is a financial instrument through which a company raises finance for funding environmentally friendly projects. But if the company indulges in misleading practices of green-washing and diverts funds elsewhere or labels their projects as green, investors may lose their trust and confidence in the market. Lose of trust in the green bond market may affect its potential growth in the long run.
2. **Market credibility:** This credibility is undermined by greenwashing, which leaves the public uncertain about the validity of the projects being financed. The market might therefore slow down as potential investors, including institutional investors, become hesitant to invest.
3. **Diversion of funds:** Money that is misallocated to initiatives that are not actually sustainable due to greenwashing can happen. In addition to failing to make a substantial impact on the environment, this also takes funds away from initiatives that might actually have a good influence. Such misallocation might deter investors from really promoting sustainability and might restrain the development of the green bond market.
4. **Long-term feasibility:** Developing a sense of collective responsibility for environmental stewardship is crucial to the long-term success of the green bond market. As a result of greenwashing, companies are able to appear environmentally conscious without making any significant changes to their operations. Investors and issuers who are genuinely interested in driving sustainable change may be deterred if the market is perceived as being filled with insincere commitments.

There are several steps to address these challenges and promote the growth of the green bond market. Green bond principles are the major and one of the most popular strategies to overcome these challenges. By outlining the issuance strategy for Green Bonds, the Green Bond Principles serve as voluntary process rules that encourage transparency and disclosure while also fostering legitimacy in the growth of the green bond market. These principles facilitate investors by confirming accessibility to the information required to appraise the impact of their investment into environmentally sustainable projects. GBP recommends a transparent disclosure process for issuers, which allows various participants to gain a better understanding of Green Bonds. The following are the four indispensable elements for adhering to the GBP: 1. Use of Proceeds 2. Project Evaluation and Selection Process 3. Proceeds Management 4. Reporting (The Green Bond Principles, 2021).

1. **Use of Proceeds:** This part focuses on the purposes for which the money earned through the sale of green bonds is utilized. The categories of qualifying projects or activities to which the bond revenues will be allocated by the issuers must be specified in detail. These activities should benefit the environment, such as those that support the preservation of biodiversity, renewable energy, sustainable transportation, or energy-efficient construction. The way the money is used should be clearly related to environmental goals.

2. **Project Evaluation and Selection Procedures:** Issuers are required to set up a thorough procedure for identifying and choosing projects that fit the specified environmental objectives. This entails evaluating the projects' possible advantages, dangers, and effects on the environment. The issuer must show that the projects it has picked for financing adhere to strict environmental requirements in order for them to support sustainability objectives.

3. **Proceeds' Management:** The credibility of green bonds must be maintained by proper administration of the raised funds. The use of proceeds section defines eligible projects, and issuers should implement systems to monitor and only allocate funds to those initiatives. Investors need to know that their money is being spent for the desired purpose, therefore transparency is essential here. To guarantee a distinct separation from other financial activity, separate accounting and reporting for proceeds from green bonds is frequently advised.

4. **Reporting:** Fundamental components of green bonds include openness and constant communication with investors. The utilization of the revenues and the environmental impact of the financed projects must be disclosed in the issuers' regular reports. These reports must be precise, understandable, and give pertinent information to help evaluate the initiatives' efficacy and progress. Transparency increases stakeholder understanding of the value of their investments in promoting sustainable results and fosters investor trust.

Following these four components, issuers can ensure their bonds are aligned with Green Market Principles.

- **Types of Green Bonds**

There are currently four types of Green Bonds shown in Table 9.1

TABLE 9.1
Types of Green Bonds

Category of Bond	Meaning
Proceeds Bond	With recourse to the issuer's entire balance sheet
	Proceeds of these bonds are invested in green and environment-friendly projects only that are associated with the green bond principles. The Investor has recourse to the entire balance sheet of the issuers, which in turn reduces the risk and encourages investors to invest
Use of Proceeds Revenue Bonds	Just like proceed bonds, investors don't have recourse to the complete balance sheet items of the issuer rather it is confined to issuers' pledged cash flow of revenues. Again, the proceeds of such bonds are only invested in green bonds that are associated with green bond principles
Green Project Bond	In the case of green project bonds, the investors/ lender has recourse on assets/projects only, wherein proceeds have been invested.
	Proceeds of such bonds can be invested in one or a group of green projects aligned with the green bond principles.
Green Securitized Bond	The bond is collateralized by a single or group of income-generating green projects.
	Revenue generated from the project is used to pay back the bond e.g., repayment of loan against the rooftop solar package and recourse is confined to the collateralized asset only.

Since the issuance of green bonds, the year 2021 has been marked as the strongest year in terms of the issuance of Green, Sustainable and Social (GSS) bonds. Green bonds amounting to $7.5 billion have been issued in the year 2021 as against $4.28 billion, 0.70 billion, 3.14 billion, 1.09 billion raised in between 2017 and 2020. Following a decline in debt issuance in 2020, Indian GSS issuance increased more than sixfold (+585%) to US$ 7.5 billion in 2021. In the last two years, the cumulative volume for GSS bonds has nearly doubled to US$ 19.5bn (Climate Bonds Initiatives, 2021). India is the sixth largest GSS market in the APAC region and 19th globally, ranking behind China, Japan, South Korea, Australia, and Singapore, with US$ 19.5 billion in cumulative issuance.

In the Indian market, USD is the most favored currency for issuing GSS debt. The 37 of the 75 issues till date and 87% of the total issuance of green bonds are in US dollars. Around the world, the preferred currency for the issuance of green bonds is the Euro, followed by US$. Ten deals totaling US$ 6.8 billion were issued in this currency only in the year 2021.

The total size of the green bonds market in India as of March 12, 2021, is $18.3, issued through 72 issuers in 3 currencies (dollar, Indian Rupee, Euro). In the year 2021 only, the total size of green-labeled bonds issuance via 24 deals for US$ 7.0 billion out of a total US$ 18.3 billion has been a remarkable portion since the inception of the market. During the year Adani Green Energy has issued green bonds amounting US$ 2.1billion. the issue was the biggest issue of green bonds ever in India.

The average deal size of green bonds issued in India was US$ 290.8 million as against US$ 125 million globally. The size of the Green bond market globally has a recorded benchmark of US$ 582.4 billion in 2021, which further declined to US$ 487.1 billion in the year 2022 indicating an absolute change of 16% during the stated time. Issuers numbering 976 from around 62 countries have floated green bonds in 35 currencies till the year 2021 (CBI, 2022).

The government of India issued its first tranche of first sovereign green bonds for around INR 80 million (Rs 4,000 crore) for five years at a coupon rate of 7.1% in January 2023 and other lots of the same denomination for ten years at coupon rates of 7.29% in February towards infusing funds in green infrastructure projects of public sector and reducing carbon intensity of the economy India's sovereign green bonds demonstrate the country's commitment to increasing renewable energy production and reducing carbon emissions through increased expenditures on renewable energy and electrification of transport

9.10   CONCLUDING REMARKS AND THE WAY FORWARD

The climate goals of India are a race against time, and greening all finance has become a priority. In order to achieve this, officials, regulators, and participants in the financial sector need to adopt a cohesive approach, make concrete efforts, and share a common vision. To accomplish all of this, the socioeconomic factors of the country must be kept in check. A paradigm shift in narratives and strong attention to sustainable financing are key to enriching the dialog at the uppermost levels. With India's noteworthy commitment to the Paris Agreement as well as the SDGs program, more information will be solicited for its corporate objectives aligning with national ones in the near future. Therefore, there is a strong need to develop unified approaches to ESG investments, green guidelines, and financial products, as well as to state the roles of the private sector, the public sector, banks, and asset managers. Besides promoting green finance, this will also fuel the country's engine of sustainable growth.

In order to finance the progression of the country to a low-carbon or carbon-neutral economy, green bonds have been popularized. By using green bonds, public and private financing flows can reach new levels of climate awareness. Investors are becoming more and more accustomed to these new green financial instruments as a result of the stunning progress in the green bond market in recent years.

Based on the findings of the paper, Global CO_2 emissions from fossil fuels and land use change have continually increased over the years till they reached their peak in 2016 and later declined and were recorded at 41.06 billion tonnes in the year 2021. As the Indian economy is growing at a continuous pace, the development of projects and increased investment into these projects have accounted for around 41% of India's GHG emissions in 2019 and are expected to increase more sharply in the coming years. By 2070 India, which is the third major emitter of carbon dioxide around the world, will need at least US$ 10 trillion to become a carbon-neutral country. The conversion of India's coal-dependent power segment to renewable energy sources will cost around US$ 8.412 trillion, while the development of capture and storage of carbon and green hydrogen technology will cost an additional $1.494 trillion (Isjwara & Ahmad, 2022). The gap of US$ 3.546 trillion between the overall investment necessary to reach net zero and the amount that domestic banks, nonbank financial institutions, and capital markets can feasibly contribute might also be filled with aid from offshore funds. Foreign investors are keen to invest in green bonds issued by developing nations like India because of their relatively favorable value and promising economic growth prospects (Singh & Sidhu, 2021). Though the benefit of low borrowing in green bond issuance till 2021 can be offset by increased interest rates in coming years, still moderate hike in interest rate may not likely affect much the interest of issuers, investors, and intermediates involved in the issuance of these bonds.

In the year 2023, Even the government of India has raised around Rs 16,000 crores through SGBs to finance the public sector for moving towards a low-carbon economy. Still, reducing carbon emissions and making Indian economy carbon neutral is a long journey and will take years and the support of investors and financial institutions to achieve it.

9.11 LIMITATIONS

The research is based on empirical facts and figures collected from reviews of literature and reports published by various departments at different times. As the green bond market is at its infancy stage to date, very few studies have been undertaken regarding bonds in the Indian context, and data available on websites had been in scattered form and was not transparent and uniform, so certain data and facts may still remain unfolded.

9.12 FUTURE RESEARCH DIRECTIONS

India International Exchange Limited (India INX), was established on January 9, 2017. The electronic platform offers trading in a variety of financial products for 22 hours daily. According to the ICMA's GBPs and Climate Bonds Initiative, India INX also has a dedicated green listing platform that offers a perfect investment environment for international investors. In line with other international listing locations like London, Luxembourg, Singapore, etc., the platform provides a debt listing mechanism. So, a detailed study about the functioning of India INX and issuers floating green debentures through this platform for raising funds from Indian and offshore clients can be an important area of future research.

REFERENCES

Bagaria, R. (2020). *Everything you need to know about green bonds in India*. Retrieved from Green Clean Guide: https://greencleanguide.com/everything-you-need-to-know-about-green-bonds-in-india/

Bansal, A. (2022). OPINION: Green financing in India — Need, significance, urgency, and way forward. *Sustainability*. Retrieved from https://www.businessinsider.in/sustainability/

article/opinion-green-financing-in-india-need-significance-urgency-and-way-forward/articleshow/92948265.cms

Baulkaran, V. (2019). Stock market reaction to green bond issuance. *Journal of Asset Management*, 331–340.

BloombergNEF report. (2023). Retrieved from Moneycontrol.com: https://www.moneycontrol.com/news/business/markets/indian-issuers-raised-nearly-43-billion-in-green-bonds-from-january-2014-to-march-2023-10752461.html

CBI. (2022). *Sustainable debts- Global state of the market 2022.* Climate Bonds Initiatives. Retrieved from https://www.climatebonds.net/files/reports/cbi_sotm_2022_03e.pdf

Chandra, M. (2015). Environmental concerns in India: Problems and solutions. *Journal of International Business and Law*, *15*(1). Retrieved from https://scholarlycommons.law.hofstra.edu/cgi/viewcontent.cgi?article=1278&context=jibl

Corporate Social Responsibility, Sustainable Development and Non Financial Reporting- Role of Banks. (2007). Retrieved from RBI: https://rbidocs.rbi.org.in/rdocs/notification/PDFs/82186.pdf

Dubash, N. K., & Rajan, S. C. (2001). The politics of power sector reform in India. *Economic and Political Weekly*, 3367–3390.

Green Bond- World Bank. (2022). Retrieved from World Bank: https://treasury.worldbank.org/en/about/unit/treasury/ibrd/ibrd-green-bonds

Green Finance in India: Progress and Challenges. (2021). Retrieved from RBI: https://www.rbi.org.in/Scripts/BS_ViewBulletin.aspx?Id=20022

Green finance: A quantitative assessment of market trends. (2022). Retrieved from thecityuk.com: https://www.thecityuk.com/media/021n0hno/green-finance-a-quantitative-assessment-of-market-trends.pdf

Gianfrate, G., & Peri, M. (2019). The green advantage: Exploring the convenience of issuing green bonds. *Journal of Cleaner Production*, *219*(6), 127–135. Retrieved from 10.1016/j.jclepro.2019.02.022

(2021). *India sustainable debt state of the market 2021.* Climate Bonds Initiatives, UK Based Monitoring Agency. Retrieved 2023, from https://www.climatebonds.net/files/reports/cbi_india_sotm_2021_final.pdf

Isjwara, R., & Ahmad, R. (2022). *India sets sights on record green bond issuance entering 2022.* Retrieved from https://www.spglobal.com/marketintelligence/en/news-insights/latest-news-headlines/india-sets-sights-on-record-green-bond-issuance-entering-2022-67940627

Jain, S. (2020). *Financing India's Green Transition.* Retrieved from https://www.orfonline.org/: https://www.orfonline.org/wp-content/uploads/2020/01/ORF_IssueBrief_338_FinancingGreenTransition_NEW27Jan.pdf

Khanna, N. et al. (2022). Landscape of Green Finance in India 2022. Retrieved from Climate Policy Initiatives: https://www.climatepolicyinitiative.org/publication/landscape-of-green-finance-in-india-2022/

Kim, M. (2015). Going green: considerations for Green Bonds issuers. *Government Finance Review*, *6*(31), 14–18.

Mecu, A. N., Chiţu, F., & Hurduzeu, G. (2021). The potential of green bond markets as drivers of change. *The Romanian Economic Journal*, *24*(82), 67–79. 10.24818/REJ/2021/82/06

Osadume, R., & University, E. O. (2021). Impact of economic growth on carbon emissions in selected West African countries, 1980–2019. *Journal of Money and Business*, *1*(1), 8–23. 10.1108/JMB-03-2021-0002

Peri, G. G. H. (2019). The green advantage: Exploring the convenience of issuing green bonds. *Journal of Cleaner Production*, *2019*(6), 127–135.

Ritchie, H., & Roser, M. (2020). CO$_2$ emissions. *Our World in Data.* Retrieved from https://ourworldindata.org/co2-emissions

Singh, V. P., & Sidhu, G. (2021). Investment sizing India's 2070 net-zero target. *Low-Carbon Economy*. Retrieved from https://www.ceew.in/cef/solutions-factory/publications/CEEW-CEF-Investment-Sizing-India%E2%80%99s-2070-Net-Zero-Target.pdf

Sartzetakis, E. S. (2020). Green bonds as an instrument to finance low carbon transition. *Economic Change and Restructuring*, *54*(4). 10.1007/s10644-020-09266-9

Stojanović, D. (2020). Green bonds as an instrument for financing renewable energy projects. 111–119.

Sultana, N. (2023). How climate change can impact GDP and jobs. *Big Story of the Day*. Retrieved from https://www.forbesindia.com/article/take-one-big-story-of-the-day/how-climate-change-can-impact-gdp-and-jobs/87673/1

The Economic Times. (2023). India's green bond issuances just 3.8 pc of overall domestic corporate bond market. Retrieved from https://economictimes.indiatimes.com/markets/bonds/indias-green-bond-issuances-just-3-8-pc-of-overall-domestic-corporate-bond-market-report/articleshow/97806148.cms?from=mdr

The Green Bond Principles. (2021). Retrieved from https://www.icmagroup.org/assets/documents/Sustainable-finance/2021-updates/Green-Bond-Principles-June-2021-100621.pdf

World Economic Statistics. (2015). Retrieved from The World Bank: http://www.worldbank.org

WorldBank. (2014). *India: Green growth – Overcoming environment challenges to promote development*. Retrieved from https://www.worldbank.org/: https://www.worldbank.org/en/news/feature/2014/03/06/green-growth-overcoming-india-environment-challenges-promote-development

10 Green Innovations Uniting Fractals and Power for Solar Panel Optimization

Senthil Kumar Natarajan and Deepak Negi

10.1 INTRODUCTION

Fractal geometry, first explored by Mandelbrot, is widely acknowledged as a universal concept in nature. These intricate patterns not only appear in physical objects and biological structures but also in transient phenomena like electrical discharge and socioeconomic networks. Due to their functional attributes, artistic fractal structures have become a subject of fundamental and applied research (Adams and Lewis, 2018; Clark and Hall, 2020). Fractal structures are used in the solar panel at the top of the layer generally used as the bus-bar comprising horizontal 'fingers' (each 50–200μm wide) intersected by several 'bars' (1–2 mm wide). See Figure 10.1 the number of fingers and bars varies between designs, optimizing their optical and electrical efficiencies. The solar industry adopted the bus-bar approach for its ability to effectively balance competing factors, outperforming simpler, single-layer designs. Fractals possess the ability to efficiently absorb light across a significant cross-section and disperse it via scattering (Allen and Turner, 2018; Garcia and Turner, 2020). Their large surface area effectively separates exactions into positive and negative charge carriers, leading to enhanced charge separation and collection at fractal interfaces (Lee and Brown, 2018; Negi et al., 2016; Wright and Patel, 2021). The interconnected network of branches in fractals facilitates the transport and concentration of charge carriers at each electrode this widespread prevalence of fractal geometry stems from the advantageous functionality it imparts to structures. Solar cells perform essential functions, including the collection of photons, their conversion into electrical charge, charge separation, and the transport and concentration of electrical charge at electrodes. To achieve a large surface area and optimal transport properties, structural patterns are connected across variable-size scales. Fractals, with their branched structures, hold promise in enhancing the efficiency of excitonic photovoltaic cells, and they have found practical applications in capacitors and antennas (Brown and Anderson, 2019; Garcia and Patel, 2018; Pickover, 1994). Theoretical calculations support the notion that fractal interfaces can improve charge separation and collection, making

DOI: 10.1201/9781003458944-10

FIGURE 10.1   Bus-bar structure solar panel.

Source: Author's own.

them promising candidates for various applications involving mass, energy, and entropy dispersal or collection across extensive spatial and temporal dimensions

10.2   PRELIMINARIES

10.2.1   MANDELBROT SET

The Mandelbrot set, a renowned mathematical fractal, is formally defined as the collection of complex plane points denoted as 'c' for which the iterative sequence governed by zn+1 = zn^2 + c, under the initial condition z0 = 0, remains confined within a bounded region. This set is named in honor of its discoverer, Benoit Mandelbrot (Noor, 2000; Smith and Johnson, 2020). The Mandelbrot set exhibits symmetrical properties concerning the x-axis within the complex plane. Notably, its intersection with the x-axis spans from-2 to 1/4. Upon closer examination, it reveals recurring patterns and intricately contoured boundaries, further unveiling a plethora of interconnected mathematical facts and esthetic allure.

10.2.2   MANDELBROT SET DEFINITION

We select the initial point 0 since it is the single critical point of the quadratic equation $\mathbb{R}_c(z) = z^k + c$, and the Mandelbrot set B is defined as the set of all $c \in \mathbb{C}$

for   which   the   orbit   of   point   0   is   bounded,   that   is, $B = \{c \in \mathbb{C}: \{\mathbb{R}_c{}^k(0)\}; k = 0, 1, 2, 3...$ *bounded*$\}$ an analogous formulation is $B = \{c \in \mathbb{C}\{\mathbb{R}_c{}^k(0)$ *does not tends to* ∞ *as n* $\rightarrow \infty\}\}$ (Patel and Johnson, 2019).

10.2.3   JULIA SET

The Julia set, situated within the complex plane, is a fractal set defined based on the behavior of a function applied to complex number inputs. Its primary purpose is to represent the set of input values for which the resultant outputs either tend toward infinity or remain within bounded limits. Within the Julia set, there are two fundamental classifications involved first is connected and second one is disconnected (Negi et al., 2016). Connected subsets come from the value of 'c' which is selected from inside the Mandelbrot set, and other hand disconnected subsets come from values of 'c' come from outside the boundary of the Mandelbrot set. Furthermore, the Julia set is classified as the collection of points where the family of iterates {Fn} fails to exhibit the properties of a normal family. Points within Julia set J(F) possess orbits that show sensitivity to initial conditions (Noor, 2000). An intensive relationship exists between the geometric characteristics of the Mandelbrot set at a specific point and the structural attributes of the corresponding Julia set. We also say that the Mandelbrot set serves as an index for navigating the Julia set's intricate structure.

10.2.4   JULIA SET DEFINITION

If $f: \mathbb{C} \rightarrow \mathbb{C}$ is a polynomial function with complex values, then the filled Julia set Q is

$$K(Q) = \{z \in \mathbb{C}: |Q^k(z)|_{k=1}^{\infty} \text{ does not tend } to \infty, \text{ as } k \rightarrow \infty\},$$

where complicated space is $\mathbb{C}$ and $Q^k(z)$ is k^{th} iterate of Q the filled Julia set's boundary, ∂KQ, is referred to as the 'Julia set' (Noor, 2000)

10.2.5   FEEDBACK PROCESS

The feedback process represents a bidirectional communication involving the exchange of feedback. In this context, it is crucial to provide feedback promptly and with consideration, avoiding delays that might lead to frustration. Timely and considerate delivery of feedback is essential, as postponing it can lead to growing dissatisfaction (Peitgen et al., 2004; Turner and Martinez, 2019). In the context of generating Julia and Mandelbrot sets, Mann and Ishikawa's orbits illustrate two-step and three-step feedback mechanisms, respectively. Furthermore, the Noor orbit represents a three-step iterative procedure defined by the function $F(z) = \sqrt{z^3} + cz^2 + 1$, which can be effectively employed to produce novel Julia and Mandelbrot sets. These orbits exemplify the utilization of feedback processes in the creation of mathematical sets (Smith and Johnson, 2020).

10.2.6 NOOR ORBIT DEFINITION

Let us, take into consideration an iteration sequence $\{x_n\}$ for the starting point $x_0 \in X$, such that the question is

$$\begin{cases} x_{k+1}: x_{k+1} = (1 - \delta)x_k + \delta_k Ty_k; \\ y_k = \left(1 - \phi_k\right)x_k + \phi_k Tz_k; \\ z_z = (1 - \varphi)x_k + \varphi_k Tx_k; \quad k = 0, 1 \dots \end{cases}$$

The sequences away from 0 and converge $\delta_k, \phi_k, \varphi_k \in [0, 1]$ and $\{\delta_k\}, \{\phi_k\}, \{\varphi_k\}$. The aforementioned repetitions are known as the Noor orbit, which is characterized by NO, a function of five tuples $(T, x_0, \delta_k, \phi_k, \varphi_k)$ (Peitgen et al., 2004).

10.3 PURPOSED METHODS

Fractals, characterized by their inherent self-replicating and self-similar properties, present a unique opportunity for the creation of surface textures that exhibit irregularities and roughness across multiple scales. The incorporation of a multicorn-like structure within the solar panel serves to augment the surface area, consequently elevating the absorption rate of incident sunlight. This heightened absorption rate, in turn, translates to increased electric current generation and output power. Such a design significantly enhances the overall efficiency and performance of the solar panel system. Through the integration of fractals into the design of surface textures for solar panels, we can introduce intricate patterns that substantially augment the active surface area available for light absorption. The utilization of the Noor formula to generate a multicorn fractal structure at a power of 5, incorporating three parameters: alpha and beta set at 0.01, while gamma is set at 0.1, results in an image structure that significantly contributes to the entrapment and guidance of incident light rays toward the active regions of the solar cell. Solar panels equipped with an increased number of antennas effectively absorb a higher quantity of photons from the incident light, enhancing the efficiency of the panel, as depicted in Figure 10.2 (Peitgen et al., 2004). The repetitive nature of the multicorn fractal pattern across various scales substantially augments the surface area of the solar cell. This expanded surface area facilitates the enhanced capture of incident light, thereby amplifying the panel's light-absorbing capabilities.

The solar panel has seven layers see Figure 10.3. Each layer is important and has its own essential work to capture the maximum light from the environment and convert it into electricity. The second layer is equipped with a specialized coating designed to minimize the reflection of incident light, thus enhancing light absorption (Carter and Hughes, 2019; Martin and Baker, 2017; White and Davis, 2017). The third layer, featuring solar cells, is primarily responsible for converting absorbed light into electricity. In this layer, a multicorn fractal structure, derived from the Noor iteration formula, is employed in conjunction with semiconductor materials to facilitate this conversion process (Rodriguez and Martinez, 2020; Taylor and King, 2017; Turner and Wilson, 2019).

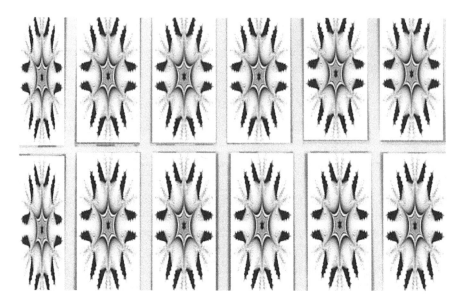

FIGURE 10.2   Multicorn fractal structure solar panel.

Source: Author's own.

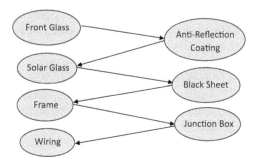

FIGURE 10.3   Seven layers of a solar panel.

Source: Author's own.

10.3.1   Generation Algorithm for Fractal Structure

Algorithm: Fractal-Based Iteration

Input:

- Initialize parameters:
  - alpha = 0.01
  - beta = 0.01
  - gama = 1
  - power = 5

- n = 100 (Number of iterations)
- c = complex(−1.6163, 0.0006) (Complex constant)

Initialization:

- Initialize arrays:
  - t[n+1] (Array to store time steps)
  - x[n+1] (Array to store complex values at each iteration)
  - absx[n+1] (Array to store absolute values of x at each iteration)
- Initialize x (Adams and Lewis, 2018) with a complex value (x0)
- Set z = complex(0, 0)

Iteration:

- For i = 1 to n:
  - Set t[i] = i − 1
  - Set x[i] = z
  - Calculate z[i] = (1 − gama) * x[i] + gama * (conj(x[i]^power + c))
  - Calculate y[i] = (1 − beta) * x[i] + beta * (conj(z[i]^power + c))
  - Calculate x[i+1] = (1 − alpha) * x[i] + alpha * (conj(y[i]^power + c))
  - Update z = x[i+1]
  - Calculate absx[i] = abs(x[i])

This algorithm represents the code's logic step by step, demonstrating how complex numbers are iteratively updated using the specified parameters, resulting in a multicorn fractal-like pattern.

10.4 CHALLENGES AND FUTURE DIRECTIONS

Complex solar panel designs that need specialized manufacturing techniques can raise production costs and have an influence on scalability, which adds to the challenges facing manufacturers (Chen and Li, 2021; Negi and Negi, 2016). These issues must be resolved in order to maximize the efficiency of fractal solar panels since they have the capacity to absorb more light from their sources.

10.5 CONCLUSION

This paper examines the optimization potential of fractals in relation to solar panels. It emphasizes the Noor iteration approach in particular and how it might produce multicorn pictures. A method for generating Multicorn fractal structures is also included in the study. This approach effectively generates complex fractal patterns appropriate for solar panel designs for the research. This method's fractal structure, which includes many antennas, makes it possible to maximize the amount of solar energy absorbed. In order to maximize efficiency, solar panels are made to catch sunlight over a bigger surface area. This methodology uses the inherent capacity of natural light.

REFERENCES

Adams, K. M., & Lewis, T. R. (2018). A comparative study of fractal and conventional solar panel designs. Journal of Green Engineering, 6(1), 45–59.

Allen, G. R., & Turner, M. J. (2018). Fractal-inspired designs for solar panel structures: A review of recent developments. Renewable Energy, 20(4), 112–127.

Brown, L. K., & Anderson, R. D. (2019). The role of fractal geometry in solar panel optimization. Solar Energy, 42(2), 201–214.

Carter, P. B., & Hughes, L. S. (2019). Fractal-inspired solar panel optimization for urban environments. Sustainable Cities and Society, 8(3), 112–127.

Chen, Q., & Li, W. (2021). Exploring the potential of fractal geometry in solar panel design. Energy Conversion and Management, 25(6), 420–435.

Clark, J. D., & Hall, A. W. (2020). Exploring fractal patterns for improved light capture in solar panels. Journal of Applied Physics, 48(2), 301–315.

Garcia, J. A., & Turner, D. L. (2020). Harnessing fractal geometry for solar panel innovation: Challenges and prospects. Solar Energy Focus, 4(2), 301–315.

Garcia, M. S., & Patel, A. (2018). Innovative fractal-based electrode patterns for enhancing solar panel performance. International Journal of Green Energy, 10(4), 301–315.

Lee, S., & Brown, M. (2018). Integrating fractal-based designs for efficient solar energy conversion. Energy. Sustainability and Society, 12(3), 45–59.

Martin, L. H., & Baker, M. A. (2017). Fractal-based approaches to improving solar panel efficiency: A comprehensive review. Journal of Sustainable Energy Engineering, 5(2), 112–127.

Negi, D., & Negi, A. (2016). A behavior of tricorns and multicorns in N-Orbit. International Journal of Applied Engineering Research, 11(1), 675–680.

Negi, D., Negi, A., & Agarwal, S. (2016). The complex key cryptosystem. International Journal of Applied Engineering Research, ISSN, 0973–4562.

Noor, M (2000). New approximation schemes for general variational inequalities. J. Math. Anal. Appl., (251), 217–229.

Patel, S., & Johnson, D. (2019). Harnessing the power of fractal geometry for solar panel enhancement. Solar Energy Materials and Solar Cells, 28(4), 301–315.

Peitgen H, Jürgens H, Saupe D. (2004). Chaos and Fractals: New Frontiers of Science. New York: Springer-Verlag.

Pickover. C (1994). Computers, Pattern, Chaos, and Beauty. New York: St. Martin's Press, 1990 Comput Graphics; 18(2), 239–248

Rodriguez, A., & Martinez, C. (2020). Advancements in fractal-inspired solar panel technology. Sustainable Energy Technologies and Assessments, 8(3), 112–127.

Smith, J. A., & Johnson, B. C. (2020). Fractal-inspired designs for improved solar panel efficiency. Journal of Renewable Energy, 15(3), 245–258.

Taylor, K. M., & King, R. S. (2017). Optimizing solar panel designs with fractal-based algorithms. Solar Energy Research & Applications, 25(4), 112–127.

Turner, A. J., & Martinez, E. C. (2019). The application of fractal geometry in solar panel manufacturing: A case study. International Journal of Advanced Manufacturing Technology, 28(4), 678–692.

Turner, H. R., & Wilson, L. S. (2019). Fractal-based algorithms for solar panel optimization. Solar Power Journal, 18(2), 112–127.

White, E. B., & Davis, P. G. (2017). Fractals and their applications in solar energy systems: A review. Renewable and Sustainable Energy Reviews, 12(5), 1234–1250.

Wright, A. D., & Patel, N. R. (2021). Designing next-generation solar panels using fractal geometry. Journal of Solar Energy Engineering, 7(2), 301–315.

11 Digital Twins in Green Manufacturing
Enhancing Sustainability and Efficiency

A. Mansurali, T. Praveen Kumar, and Ramakrishnan Swamynathan

11.1 INTRODUCTION

With greater concern for the environment and demand for more accountable production practices, the concept of green manufacturing, which also refers to sustainable products, has gained considerable attention recently. In order to minimise the environmental impact of production activities, while also supporting resource conservation and reducing waste, green manufacturing refers to adopting environmentally friendly practices, technologies or strategies. Major environmental challenges, such as air and water pollution, deforestation and greenhouse gas emissions have been created by the Industrial Revolution and subsequent rapid industrialisation. The Intergovernmental Panel on Climate Change (IPCC) reports that human activities, including manufacturing, are major contributors to greenhouse gas emissions, leading to global warming and climate change (IPCC, 2014). These findings have led to increased calls for sustainable practices that may mitigate the environmental impact of manufacturing activities. Traditional production processes rely significantly on unavailable, finite resources such as fossil fuels and minerals which are subject to depletion. The World Economic Forum is warning that, by 2050, world demand for resources may be more than the Earth's ability if current patterns of resource consumption continue. The urgency to adopt production practices that conserve resources and keep them available for the next generation is underlined in this scenario. In order to combat pollution and support good manufacturing practice, governments all over the world are implementing more stringent environmental legislation. For instance, the European Union's Eco-design Directive sets requirements for energy-efficient design in manufacturing processes and products (European Commission, 2009). Moreover, consumer preferences for products that are environmentally friendly will also shift to greener choices and companies adopting sustainable production practices have an edge over competitors on a global scale by virtue of this trend (Nielsen, 2015). Significant amounts of waste and resource depletion have been generated in the Traditional

DOI: 10.1201/9781003458944-11

Linear Economy, where products are produced, consumed or discarded. On the other hand, by encouraging recycling, reutilisation and reprocessing of products with a view to reducing waste and extending their product lifetimes, the Circular Economy model aims at closing the loop. In the transition to a circular economy, green manufacturing is an essential component (Ellen MacArthur Foundation, 2012). The economic benefits to enterprises, such as cost savings through energy efficiency and waste reduction, are offered by sustainable manufacturing practices. In addition, green manufacturing practices are in line with CSR initiatives by companies and contribute to enhancing their reputation and social impact (Delmas & Toffel, 2004). The environmental challenges, resource depletion, regulatory pressures, consumer demand, the principles of a circular economy and economic advantages justify an urgent need for green manufacturing. Fostering a more efficient and socially responsible industrial sector requires the adoption of sustainable production practices, not only to mitigate environmental risks but also to promote a more efficient and socially responsible industrial sector. This chapter aims to Construct a Digital twin implementation framework and address the challenges and considerations associated with the implementation of digital twins within green production. The structure of the paper is as follows: Section 11.1 is Introduction Green Manufacturing: Overview and Sustainable Practices: Need. Section 11.2 is Review of Literature Digital twins: Overview and significance. Section 11.3 is Methodology Design and Framework for implementing digital twins Section 11.4: Analysing the Challenges and how they're overcome in Implementing digital twins Section 11.5: Future Direction, Implication and Conclusion.

11.2 REVIEW OF LITERATURE

11.2.1 DIGITAL TWINS: OVERVIEW AND ITS SIGNIFICANCE ACROSS SECTORS

A cutting-edge concept that involves the creation of a virtual representation of a physical object, process or system is digital twin technology. Using the data from sensors, Internet of Things equipment and others, this virtual replica is continuously updated in real time as a digital twin. Digital twin technology is offering significant insight, predictive capacities and optimisation opportunities for a broad range of industry sectors through its link to the Physical and Digital worlds. Its relevance is based on its capacity to improve operational efficiency, develop new products and streamline the way decisions are made. Several key components are involved in digital twins, e.g., sensors for collection of information, data insertion and analysis platforms as well as visual interfaces. Realtime synchronisation, bipartite communication and the ability for accurate simulation and prediction of behaviour are characteristics of this technology.

By providing real-time information on machine efficiency, production processes and quality of the product, digital twins play an important role in smart manufacturing. It helps manufacturers to optimise production, decrease interruptions and improve overall efficiency (Tayo et al., 2018). Digital twins are used to monitor the health and performance of aircraft components, engines and systems in the Aviation sector. They're supporting predictive maintenance, which reduces operational costs and enhances security (Raschecker et al., 2017). Power generation companies are helping to manage electricity generation, distribution and consumption more effectively through digital twins. They help to monitor equipment condition, optimise energy consumption and predict maintenance needs (Trommer et al., 2021). Digital twins are being employed in the medical field to model patient characteristics, drug development simulations and device testing. It allows patients to have a personalised treatment plan and improve their outcomes(Viceconti et al., 2020; Rakshit & Sharma, 2021). By simulating infrastructure, transport systems and energy grids, digital twins are contributing to smart city projects. They're helping urban planning authorities make informed decisions on sustainable development, (Neophytou et al., 2019).

The digital twins are facilitating vehicle design and testing, making it easier for manufacturers to achieve better performance, security or fuel efficiency. They help to predict maintenance and improve the driving experience, according to Gericke et al., 2018. For buildings and infrastructure, digital twins help to plan construction projects, monitor progress and predict maintenance. They are designed to increase efficiency and reduce risks related to construction, (Mensah et al., 2020; Rajesh et al., 2019). Digital twin technology, which provides an innovative approach for optimising processes, enhancing decision making and improving performance, is relevant in many sectors. Digital twins are helping to advance the fourth industrial revolution by taking advantage of real-time data and superior simulations, creating innovation and efficiency in a variety of sectors.

11.2.2 Types of Digital Twins

Product digital twins are virtual copies of real products, e.g., machines, vehicles, consumer goods or any other object in its own right. These digital twins play a role in the design, simulation and optimisation of products. In order to improve product performance, perform virtual tests and enable predictive maintenance, they are used in industries such as automotive, aerospace and consumer electronics (Li et al., 2019).

Process digital twins are models and simulations of manufacturing or industrial processes. They have been designed to optimise the production process, identify bottlenecks and reduce interruptions. Process digital twins are used to assist in process improvement, resource utilisation and quality control for sectors such as production, chemical or energy. (Tao et al., 2019).

System digital twins help in urban planning, energy management and optimising the performance of large-scale infrastructure. For example, energy grids, smart cities and transportation networks are complex systems or infrastructures that can be classified as System Digital Twins. Analysis and optimisation of system-level performance can be carried out by these Digital Twins. (Wang et al., 2021).

Performance Digital twins are devoted to monitoring and optimising the performance of specific components or subsystems in the system as a whole. For monitoring the health and performance of single components, e.g., medical devices or aircraft engines, they have been applied extensively in sectors like healthcare and aviation (Huang et al., 2021).

Human interaction with a virtual model, enabling simulations and predictions based on human decisions and behaviour, is part of the Human loop of digital twins. In areas such as health, where digital twins support personalised treatment planning and decisions based on the particular characteristics of patients (Viceconti et al., 2020).

11.2.3 ROLE OF DIGITAL TWINS IN MANUFACTURING

By providing data-driven insights and opportunities for optimisation that support sustainability, resource efficiency and reduction of the impact on the environment, digital twins are playing an essential role in helping achieve Green Manufacturing's objectives. Real-time monitoring, prognostic analysis and resource optimisation resulting in improved environmental performance and operational efficiency are made possible through the integration of digital twin technology into green manufacturing processes (Rana et al., 2022).

Digital twins simulate the impact of product and material designs on the environment, thus promoting sustainable design practices. Through virtual proto-typing and analysis, manufacturers can identify energy-efficient design alternatives and assess the product's life cycle environmental footprint. By optimising at an early stage in its development, this approach contributes to reducing materials waste and energy consumption over the product's life cycle. Digital twins are identifying opportunities for energy optimisation and resource efficiency through continuous tracking of manufacturing processes. Manufacturers can optimise production parameters and cut their energy consumption and greenhouse gas emissions by using continuous data from sensors and Internet of Things devices. This ability will help achieve the energy savings targets and reduce carbon footprints in manufac-turing operations, (Xiao et al., 2021).

Digital twins are an instrument that allows for real-time monitoring and analysis of emissions in the manufacturing process. In order to fulfil regulatory requirements and reduce harmful pollutants, producers may take proactive measures by predicting emissions and impacts on the environment. In order to align manufacturing practices with the Sustainable Development Goals, this aspect is crucial(Schmitt et al., 2020). Predictive maintenance procedures contributing to energy efficiency are supported by digital twins. Manufacturers will be able to anticipate equipment failure and schedule maintenance activities in the absence of peak energy consumption by analysing data from sensors and machine performance. This approach reduces unplanned blackouts and energy waste resulting in unexpected breakdowns. The Digital Twin helps to manage a supply chain by providing accurate visibility in the whole supply chain network. Manufacturers may make informed decisions to optimise their supply chains for sustainability and efficiency, by tracking the environmental effects of raw materials, transport and logistics(Tao et al., 2019).

11.3 OBJECTIVE OF THE STUDY

- To construct a digital twin implementation framework
- To address the challenges and considerations associated with the imple-mentation of digital twins within green production.

11.4 METHODOLOGY

This chapter "Digital Twins in Green Manufacturing: Enhancing Sustainability and Efficiency," uses a systematic approach that combines many research methodological components. In order to study the role of digital twins in the context of green manufacturing and to offer useful insights and recommendations for stakeholders, the research design for this chapter is primarily exploratory and descriptive. Using case studies to demonstrate the advantages of digital twins in actual green manufacturing scenarios and quantitative data analysis to measure the effect of digital twins on important sustainability and efficiency metrics, the research type combines qualitative and quantitative analysis. The research approach includes a thorough literature analysis to lay the groundwork for understanding digital twins, sustainable manufacturing and green manufacturing in the manufacturing industry. The conceptual framework, which serves as the foundation for comprehending the connections between digital twins, sustainability and efficiency, is informed by this review. The sampling strategy focuses on choosing representative case studies from various green manufacturing industries to show how digital twins can be used in various settings. These case studies will offer in-depth qualitative insights into the implementation, difficulties and results of the use of digital twins in environmentally friendly manufacturing processes. For this study, pertinent data were collected from the case studies that were chosen as well as secondary sources including academic journals and industry reports. The analytical strategy includes a comparison analysis, which compares the effectiveness of manufacturing processes with digital twin integration to those without it. This analysis tries to measure the enhancements in resource optimisation, waste reduction and energy efficiency made possible by digital twins.

11.4.1 Digital Twin Architecture

To ensure seamless integration of subsystems and to allow data processing and analysis to work effectively, the Digital Twin architecture has been designed. In most cases, the architecture is structured in layers as it looks in Figure 11.1 shows the smooth flow of data.

The perception layer collects real-time data by using sensors and IoT devices from the physical objects. The network layer acts as a communication facilitator between the perception and data integration layer for the smooth transmission of data from sensors to digital twins. The data integration layer takes care of the data integration and processing from various data sources (Sensors, IoT and databases), it checks the data readiness for analysis. The analytics layer performs analysis by using machine learning algorithms, providing insights about the data and predicting the physical object's behaviour. The application layer contains a graphical user

FIGURE 11.1   Digital twin architecture.

interface and additional applications that enable users to engage in the Digital Twin, visualise insights or make data-driven decisions. At the end, the physical layer represents the physical entity that has been modelled by a digital twin. It is the real-world counterpart that the digital twin mirrors.

11.4.2 IMPLEMENTATION DESIGN AND FRAMEWORK

A coherent framework, laying down critical steps and considerations for the successful implementation of digital twins in the production sector, is needed. A general framework for the application of Digital Twin in manufacturing is set out in Figure 11.2.

It's necessary to ensure that the framework mentioned in Figure 11.2 is followed for successful implementation. The following explanation provides the

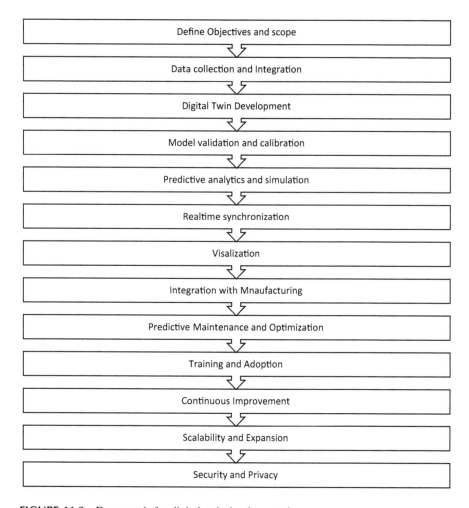

FIGURE 11.2 Framework for digital twin implementation.

understanding of a detailed framework for the application of Digital Twin in manufacturing

Step 1: The objectives of digital twinning in the manufacturing sector need to be clearly defined. To specify the particular processes, equipment or systems that will be covered by digital twins. Clear objectives such as increasing the efficiency of processes, predicting maintenance and optimising resources should be set.

Step 2: Determine data requirements to create a precise digital twin. Determine the sensors, Internet of Things devices and data sources that are necessary for real-time data collection. Make sure that the data are of high quality and provide mechanisms for their integration to a centralised platform.

Step 3: Using this data, build virtual models of the physical systems. Develop a digital twin architecture to ensure an accurate representation of the physical entity's behaviour and performance.

Step 4: Assess the accuracy of digital twin models by comparing their predictions with real world data. To make sure that the model is compatible with the physical system's behaviour, calibrate it as necessary.

Step 5: In the digital twin, use predictive analytics and simulation capabilities. To predict behaviour, find possible problems and optimise performance, use AI algorithms and simulation techniques.

Step 6: Make it possible to synchronise a Digital twin and physical system at the same time. Ensure that, in order to reflect the present state of a physical object, the Digital Twin shall be provided with real-time sensor updates.

Step 7: Develop a user-friendly graphical interface allowing users to interact with the digital twin. Help decision-makers make decisions more easily by providing a clear understanding of insight, visualisations and simulations.

Step 8: Integrate the digital twin into current manufacturing systems, By using MES (manufacturing execution systems) and SCADA(supervisory control and data acquisition). Ensure that data are exchanged and communicated in a seamless manner.

Step 9: To predict maintenance and optimise, use a digital twin. In order to reduce downtime and increase resource effectiveness, adopt proactive maintenance strategies based on the information gained from the Digital Twin.

Step 10: Train the personnel on how to use Digital Twins and its benefits. In order to ensure that the value of Digital Twin implementation is maximised, encourage mutual adoption and collaboration between different departments and teams.

Step 11: The digital twins' performance is constantly monitored and evaluated. To find gaps for improvement and refinement, gather comments from users and stakeholders.

Step 12: Sufficiently scale up the implementation of a digital twin, covering other processes, equipment or systems as necessary.

Step 13: To protect a digital twin and the data it creates, adopt robust cyber security measures. To ensure the security of sensitive information, in compliance with data protection legislation.

11.4.3   Implementation Strategies: Digital Twins in Manufacturing

Collaboration between stakeholders and industry initiatives is often the key to successfully adopting digital twins in green manufacturing. Industry can help to speed up the adoption of Digital Twins and support sustainable practices through the dissemination of knowledge, good practice and standardisation efforts. The various cooperation methods of industrywide involvement are listed below:

11.4.3.1   Collaborative Research with Academia and Industry

The development and validation of Digital Twin Technology in the area of Green Production can be accelerated through collaboration between academia, research institutions and industry. Consideration of common challenges and insight into innovative applications can be ensured through joint research projects.

11.4.3.2   Collaborative Industry Frameworks and Standards

The development of joint frameworks and interoperability standards for digital twins can be facilitated through participation in Industry Consortiums and Standards Organisations. Industry can develop best practices and guidelines for the implementation of Digital Twins in an economical manner by cooperating.

11.4.3.3   PPP Model

In order to support the adoption of Digital Twin in green manufacturing, PPPs could play an important role. Funding opportunities, legal assistance and knowledge sharing may be fostered through cooperation with government agencies and industry associations.

11.4.3.4   Organising Events

Awareness of the Digital Twin and its benefits in green manufacturing can be promoted through workshops, seminars or events that disseminate knowledge. In order to facilitate a collaborative learning environment, firms can share experiences, success stories and challenges.

11.4.3.5   Pilot Projects

The feasibility and benefits of digital twins in the production of renewable energy can be demonstrated by joint pilot projects between several stakeholders from different sectors. These projects can show the world how similar practices are being implemented and inspire more companies to do so.

11.5 DISCUSSION

There are many challenges and obstacles to the implementation of Digital Twins for Green Production. The full realisation of the potential benefits of digital twins for improving sustainability and resource efficiency requires overcoming these obstacles. These common problems and strategies for addressing them are listed below:

Problem 1: It may be difficult and problematic to integrate data from different sources, ensure that they are of the right quality or manage a large amount of Realtime data.

Strategy 1: In order to ensure the accuracy and consistency of data, implement Data Governance Practices, Establish Data Standards or Invest in Strong Data Integration and Management Systems.

Problem 2: Substantial initial investment in hardware, software and human resources may be required for the implementation of digital twins.

Strategy 2: Before undertaking full-scale investments, companies may start pilot projects or phases of implementation to demonstrate benefits. The sharing of costs can also be facilitated by cooperation with technology providers and research institutions.

Problem 3: The challenge can be to ensure interoperability and compatibility between existing production systems, Internet of Things devices and Digital Twin Platforms.

Strategy 3: To ensure complete communication and exchange of data among various systems and components, use open standards and protocols

Problem 4: Particular expertise in data analysis, modelling and domain specificity is needed for the development and management of digital twins.

Strategy 4: Investment in training towards the employees' upskilling.

Problem 5: It may be a problem to deal with sensitive data from the manufacturer and ensure its security and privacy.

Strategy 5: For the protection of data, we shall implement robust security measures, encryption and access controls. Define and implement coherent data governance policies in line with the relevant regulations on Data Protection

Problem 6: It may be difficult to scale the implementation of digital twins in a number of production facilities and processes.

Strategy 6: Start to plan for scale, making sure that a digital twin architecture is elastic and capable of meeting various scenarios and requirements.

11.6 CONCLUSION

The transformative potential of digital twins is emerging as a beacon of hope to improve sustainability and efficiency in the dynamic landscape of modern manufacturing. In these Digital Replicas, we are bridging the worlds of Physical and Virtual with an extraordinary opportunity to change how the industry operates and interacts with its environment. Beyond factory floors, the influence of Digital Twin is felt in other areas. They bring about a cascade effect, from management of the supply chain to product design and energy consumption down to compliance with legislation. In line with the circular economy principles, such twins are capable of anticipating equipment failure, optimising energy use and guiding sustainable design choices. Lastly, it is a beacon of hope that digital twins will lead to greater harmony between industry and the environment by enhancing sustainability and production efficiency. The choice is ours to take advantage of this opportunity, to harness the power of technology and to create a future where sustainability and efficiency are not merely aspirations, but the basis of our manufacturing efforts, as we stand on the brink of a new era.

11.7 IMPLICATION

The integration of digital twins into green manufacturing represents a major opportunity to shape the future of sustainable production practices. Developments and advantages of digital twins have far-reaching impacts on environmental protection, resource efficiency as well as the general course of industrial production towards a more sustainable future. The potential of Digital Twins to lead the major transition towards more Sustainable, Effective and Environmentally Responsible Industrial Practices is important for the future of Green Manufacturing. Digital twins are set to play an important role in shaping a greener and more sustainable manufacturing landscape, through the use of data, simulation as well as predictive analytics.

11.8 FUTURE DIRECTION AND OPPORTUNITIES

The potential to drive sustainability and resource efficiency in the green manufacturing of is growing more promising as Digital Twin Technology develops. A wide range of future directions and opportunities, which can further enhance environmental performance and promote sustainable practice, are opened up by integrating digital twins into Green manufacturing processes. By optimising the use of materials, supporting remanufacturing processes and enabling effective end-of-life product management, digital twins can play an important role in promoting circular economy principles. Improved digital twins may be crucial for modelling whole product life cycles, as well as guiding decisions to minimise waste and maximise resource utilisation in forthcoming developments. Technological progress in renewable energies, energy storage technologies and microgrids can make it possible for production facilities to become greener than ever before. In order to facilitate the integration of renewable energy systems, anticipate electricity generation and use patterns as well as optimising energy utilisation in connection with interconnections between production and Local Interconnected Energy Networks, it would also be possible to implement cyber twins. The ability to predict digital twins can be improved by the advancement of AI and machine learning. Hidden opportunities to save energy, reduce waste and optimise processes can be identified through artificial intelligence algorithms that identify subtle patterns in data so as to continuously improve sustainability performance. In line with each customer's preferences and sustainability criteria, it could be possible to customise eco-friendly products by means of 2D twins. Such a concept can drive the development of customised production processes that reduce waste and energy consumption. By automatically collecting data, analysing and reporting information on environmental standards and certificates, advanced digital twins could simplify regulatory compliance. The process of auditing could therefore be simplified, ensuring compliance with sustainability requirements.

REFERENCES

Delmas, M. A., & Toffel, M. W. (2004). Stakeholders and Environmental Management Practices: An Institutional Framework. Business Strategy and the Environment, 13(4), 209–222. doi:10.1002/bse.414.

Ellen MacArthur Foundation. (2012). Towards the Circular Economy: Economic and Business Rationale for an Accelerated Transition. Ellen MacArthur Foundation.

European Commission. (2009). Directive 2009/125/EC of the European Parliament and of the Council of 21 October 2009 establishing a framework for the setting of ecodesign requirements for energy-related products. Official Journal of the European Union, L 285/10.

Gericke, K., Meboldt, M., Stark, R., & Kühnle, A. (2018). Use of Digital Twins in the Automotive Industry: A Review. Procedia Manufacturing, 16, 12–19.

Huang, Y., Hu, X., Qiao, J., Zhao, H., & Yu, H. (2021). Development and Application of Digital Twins in Performance Monitoring for Industrial Processes. Sensors, 21(6), 2195.

IPCC. (2014). Climate Change 2014: Synthesis Report. Contribution of Working Groups I, II, and III to the Fifth Assessment Report of the Intergovernmental Panel on Climate Change. IPCC.

Li, P., Yang, C., Yao, Y., Wen, X., Wang, R., & Lu, X. (2019). Review on Digital Twin and its Potential Application in Advanced Manufacturing. Frontiers in Mechanical Engineering, 5, 39.

Mensah, S., Ayarkwa, J., & Nani, G. (2020). A Theoretical Framework for Conceptualizing Contractors' Adaptation to Environmentally Sustainable Construction. International Journal of Construction Management, 20, 801–811. 10.1080/15623599.2018.1484860.

Neophytou, M., Tomonori, H., & Scherer, R. J. (2019). Building a Digital Twin City for Integrated Urban Design and Planning. Sustainability, 11(7), 2019.

Nielsen, C. (2015). The Sustainability Imperative: New Insights on Consumer Expectations. Nielsen Company New York.

Rakshit, P., & Sharma, R. (2021). A Study to Comprehend Role of Artificial Intelligence in Building Smart Cities. Engineering and Technology Journal for Research and Innovation (ETJRI) ISSN, 3(2), 2581–8678.

Rana, G, Khang, A., Sharma, R., Goel, A., & Dubey, A. (2022). Reinventing Manufacturing and Business Processes through Artificial Intelligence. (1st ed). CRC Press. 10.1201/9781003145011. ISBN: 9781003145011.

Rauschecker, A. M., Tornquist, D. C., Barksdale, C. C., Collins, E. G., Goggins, K. A., & Schmalzel, J. L. (2017). Model-Based Digital Twin for Predictive Maintenance of Aerospace Systems. Proceedings of the IEEE, 105(3), 500–520.

Schmitt, R., Frank, A., Wartzack, S., & Maier, A. (2020). Digital Twin in Green Manufacturing. Procedia CIRP, 88, 28–33.

Tao, F., Cheng, Y., Da Xu, L., Zhang, L., Li, B. H., & Hu, T. (2018). CCIoT-CMfg: Cloud Computing and Internet of Things-Based Cloud Manufacturing Service System. IEEE Transactions on Industrial Informatics, 10(2), 1435–1442.

Tao, F., Zhang, M., & Nee, A. Y. (2019). Digital Twin Driven Smart Manufacturing. Springer Singapore.

Trommer, S., Daum, M., & Wollschlaeger, M. (2021). Digital Twins for Energy Systems: A Comprehensive Survey. Applied Energy, 282, 116113.

Viceconti, M., Henney, A., Morley-Fletcher, E., Inzitari, D., & Quinn, T. (2020). Use of the Digital Twin Concept in Health Care: Scoping Review. Journal of Medical Internet Research, 22(9), e19230.

Wang, F., Wang, P., Li, W., & Zhang, X. (2021). An Improved Digital Twin Framework for Real-Time Cyber-Physical Systems in Smart Cities. Sensors, 21(5), 1660.

Xiao, J., Yuan, J., Geng, Y., Qian, Y., & Huisingh, D. (2021). Digital Twin-Enabled Green Manufacturing. Journal of Cleaner Production, 298.

12 Case Studies on Circular Economy Model
Green Innovations in Waste Management Industry

S. Senthil Kumar

12.1 INTRODUCTION

The Industrial Revolution, which started in the late 18th century, brought about significant technological advancements, manufacturing and economic growth. However, it also had profound and far-reaching adverse consequences on the environment. Since the production and consumption processes that originated from the Industrial Revolution never accounted for the cost of the environment, it led to severe ecological degradation. Some of the industrial revolution's significant environmental consequences include climate change, air, water and ocean pollution, soil contamination, and loss of biodiversity and ecosystem services.

Traditionally, businesses operated with a "Greed is Good" mindset that prioritised individual greed and profit maximisation without considering broader environmental and societal consequences (Kenway & Fahey, 2014). The traditional production and consumption processes used a linear "take-make-dispose" approach, pushing our natural ecosystem beyond its regenerative capacity. The unprecedented extraction and consumption of natural resources led to depletion, thus raising concerns about the potential of future generations to meet their needs. Since the economy is embedded in the environment, societies and governments called for positive actions from the industry to protect the environment in the early 1990s. Initially, there was resistance and scepticism from some industries regarding environmental concerns, often driven by the perception that environmental regulations would be costly and hinder economic growth. However, over the years, there has been a shift in the attitudes and actions of industries towards environmental protection. As a result of community and governmental pressures, firms have been compelled to reevaluate their production processes. There is a growing recognition that environmental protection and sustainable business practices are necessary for preserving the planet and ensuring long-term business viability and resilience in a changing world. Businesses have started pivoting from the "Greed is Good" model to the "Do Good to Do Well" model that incorporates environmental and social considerations into business strategies (Falck & Heblich, 2007).

DOI: 10.1201/9781003458944-12

Due to rapid urbanisation, population growth, and changing consumption patterns, India faces enormous challenges in managing waste. Inadequate waste management infrastructure, inefficient waste collection and segregation, and low levels of awareness towards cleanliness and recycling have increased the complexity of the problem. Although the quantity of waste generated is rising, waste collection efficiency in India is still striving to keep pace. The waste collection efficiency in metropolitan cities ranges from 70 to 90%, while many smaller cities fall below 50%.

The principles of production and consumption of the circular economy concept have emerged as promising solutions to address waste management challenges. Unlike the linear use and disposal model, the circular economy envisions a closed-loop system where materials are perpetually circulated and repurposed. The circular economy encompasses a novel approach, framework, and process, aiming to offer an alternative to conventional linear production and waste disposal systems. In this model, resources are managed to extract maximum value and minimise waste at every stage of their life cycle (Walter, 2016). Implementing the circular economy model offers India a potent solution to its waste management issues. This economic approach emphasises minimising waste generation and optimising resource utilisation by prolonging their lifecycle. Embracing the principles of "reduce, reuse, and recycle" can yield numerous benefits for waste management in the country. India can effectively curtail waste generation by designing durable products and encouraging recycling and reusing. Businesses must innovate to disruptively move from a linear economic model to a circular one. Sustained economic growth and progress are not solely dependent on capital accumulation but primarily driven by technological advancements and the introduction of innovative products and services. Such innovations are often called green since they entail creating and applying novel methods of production and consumption, organisational procedures, and management ideas aimed at sustaining the environment and lessening human activity's damaging effects on the environment.

Rapid urbanisation, economic growth, population growth and urban consumption have put India in one of the top ten countries in the world in generating municipal solid waste. According to The Energy and Resources Institute report, the country produces more than 62 million tonnes of waste annually. However, only 43 million tonnes of this waste is collected, with 12 million tonnes being treated before disposal. The remaining 31 million tonnes are discarded in waste yards without proper treatment. The inadequate waste management system has led to significant environmental and public health concerns in India. Traditionally, waste is managed by government bodies. However, government bodies lack the technical and management capacity to deal with this problem. Entrepreneurial solutions adopting circular economy principles can potentially address India's waste problem. The legal and political environments have evolved to support such entrepreneurship. Past research recognised the potential of circular business models in solving our waste management issues (Ghisellini et al., 2016).

The current study aims to study green innovations adopted by a few organisations focused on waste management in India and analyse the opportunities and challenges to progress towards a circular economy. The paper is divided into six sections. The following section describes the research methodology. The third section reviews the literature on circular economy principles and India's waste management issues. The fourth section describes the cases of Ecokaari, Graviky Labs and Padcare Labs. The fifth section analyses the opportunities and challenges in the market for green innovation, and the last section provides a conclusion.

12.2 RESEARCH METHODOLOGY

The current study uses case study research, typically used as an inquiry method in social science. Case study research is a qualitative research method that involves an in-depth and detailed examination of a specific organisation. Dul and Hak (2007) defined a case study as "an empirical inquiry that investigates a contemporary phenomenon within its real-life context, especially when the boundaries between the object of study and context are not clearly evident" (p.4). A case study strategy would be appropriate if a researcher wants to study contextual conditions. Case study research designs are flexible in utilising various evidence collected from documents, interviews, and observations. A multiple-case research study offers a more profound comprehension of individual cases within a group by examining their similarities and differences. Evidence obtained from multiple-case studies is considered more dependable than single-case research. Multiple case studies enable a more extensive investigation of research questions and aid in theory development.

12.3 LITERATURE REVIEW

The industrial revolution's negative impacts caused by human activities have given rise to the concept of sustainable development (SD). SD is built upon fairness, addressing justice for both present and future generations concerning resource utilisation. SD focuses on the equitable distribution of resources among the competing interests of multiple stakeholders. The goals of SD are to conserve nonrenewable natural resources and responsible use of renewable resources while ensuring the overall development of human beings (Borowy, 2013; Yosef, 2008). Industries rely on natural resources such as water, energy, raw materials, and ecosystem services to operate and produce goods and services. Sustainable practices ensure the responsible use and conservation of these resources, ensuring their availability in the long term.

Several approaches have been proposed to help industries balance economic growth with environmental protection.

12.3.1 CIRCULAR ECONOMY PRINCIPLES

According to a report published jointly by the Ellen MacArthur Foundation and McKinsey in 2012, embracing a circular economy (CE) leads to significant environmental and social advantages apart from substantial economic benefits (Ellen MacArthur Foundation 2012, p. 5). The notion of circular economy is characterised by restorative and regenerative designs that use products, components and materials at their optimum levels and value. The CE delinks economic development from finite resource utilisation, and it is based on three principles.

The first principle is to maintain and improve the natural capital. The need for physical materials or resources is reduced by delivering a utility or service in a virtual or non-physical form. For example, the dematerialising utility in academic publishing can be seen in the transition from physical media like books and journals to digital copies. So, a circular system advocates the usage of suitable technology and processes that use renewable or better-performing resources where possible. A circular economy also enhances natural capital by facilitating the movement of nutrients within the system and creating an environment that supports the rejuvenation of vital components, such as Earth and water resources. In urban settlements, the water bodies are rejuvenated by preventing pollution and removing the obstructions in inlet and outlet systems.

The second principle aims to maximise resource utility by continuously circulating products, components, and materials in their most valuable forms. Unlike the traditional linear "take-make-dispose" model, the circular economy aims to create a closed-loop system where resources are circulated as long as possible by designing products for durability, repairability, and disassembly and implementing strategies to extend their lifespan and extract maximum value from them. The goal of continuous circulation is to minimise resource consumption and reduce waste generation. This approach promotes the efficient use of resources, encourages the development of sustainable business models, and fosters innovation in design, manufacturing, and recycling processes. For example, bio-degradable waste can be processed to break down into nutrient-rich compost. The composted material provides essential nutrients to the soil, thus reducing the need for synthetic fertilisers. Agricultural waste is processed to manufacture compostable cutlery items after use, thus transforming into a valuable resource.

The Circular Economy principles aim to minimise the impact of the firm's operations on the environment by a) eliminating waste and pollution, b) circulating products and materials at their highest value in the production and consumption processes, and c) shifting focus from extraction to regeneration. The current linear economic system is unsustainable due to finite resources. The linear economy produces waste and pollutes the environment beyond its regenerative capacity, but waste is considered a design flaw in a circular economy. Natural systems have regenerative capacity, and nothing is wasted in nature. The products and materials are continually circulated through maintenance, sharing, reusing, repairing, refurbishing, remanufacturing and recycling. When we move from a linear to a circular economy, we build natural capital, and our production and consumption processes will not degrade nature but help it regenerate itself (Webster, 2021; Charter, 2018).

12.3.2 India's Waste Problem

The changes in socioeconomic factors in India, such as population growth, industrialisation, and labour migration leading to rapid urbanisation, have led to a significant increase in waste generation in the country. The widely prevalent practices such as open dumping and burning solid waste and letting industrial waste into the water bodies lead to releasing harmful greenhouse gases, particulate matter and toxic substances into the atmosphere and contaminating soil and groundwater. According to the Planning Commission Report (2014), India produces about 70 million tonnes of MSW per year, and it is estimated to reach around 165 million tonnes by 2030 and potentially about 436 million tonnes by 2050. Approximately 80% of the MSW is collected, and only about 28% is effectively managed. The improper disposal of MSW is associated with unscientific practices, urbanisation, population growth, societal norms, and a lack of environmental awareness. Open dumping of MSW, which is a norm across many Indian cities and towns, has severe negative impacts on the environment and human health. The unscientific collection and inefficient transportation methods led to the massive accumulation of MSW throughout the country (Meena et al., 2023).

The circular economy model treats waste as an asset rather than a burden. Green innovations operate with reduce, reuse, and recycle principles to extend the lifespan of waste by keeping them in circulation and extracting the utmost value from them. The reduction method is widely recognised as the most valuable strategy for managing waste due to its cost-effectiveness and environmental benefits. Organisations can reduce waste through digitisation of the activities found in the value stream map, adopting GUIs in Kanban, lean automation, identification of energy waste, and lower fuel usage in transportation, interconnecting workers with managers through the cloud and cyber-physical systems, redesigning the product to lower environmental impact (Yadhav, et al., 2023). Reuse methods aim to extend the lifespan of materials and products, thus minimising the need for fresh resources. Manufacturers shall repair and refurbish, replace worn-out components, and salvage the pieces and parts from discarded products that can be reused in manufacturing new products. Other methods include redistributing excess or unsold inventory collection of durable containers and packaging materials, which can be cleaned and reused in the supply chain multiple times. NLC India Limited (NLCIL), a coal-based power generation company, generates a significant amount of fly ash as a byproduct. Fly ash is a fine powder consisting of mineral residues from coal combustion. It is used in constructing embankments, as a raw material for manufacturing building materials like bricks, blocks, cement, and land reclamation projects. Waste is sorted and separated into different recyclable materials, such as paper, plastics, metals and glasses, using material recovery facilities. Paper is recycled into tissue paper, packaging materials, and more. Plastics are recycled into plastic pellets or flakes. Metal is recycled into cans, automotive parts, construction materials, and appliances. Glass is turned into new products such as bottles, jars, and fibreglass insulation.

Environmentally responsible businesses can be classified into two categories: green businesses and green-green businesses. The main difference between them is their origin and approach towards sustainability. A "green business" typically evolves from an existing enterprise that adopts environmentally sustainable practices after realising the cost, innovation, marketing advantages, and ethical arguments. These businesses try to reduce their environmental impact, embrace greener processes and products, and often experience financial benefits. On the other hand, a "green-green business" is intentionally designed from the start to be environmentally responsible and sustainable. It emerged as a startup that aimed to transform its industrial sector towards a model of sustainable development (Isaak, 2016). Both these businesses leverage innovation to achieve their green goals. Green innovation refers to creating, adopting, or utilising an organisation's novel products, processes, services, or management/business methods. These innovations are characterised by their ability to reduce environmental risks, pollution, and negative impacts on resource use throughout their life cycle (Kemp & Pearson, 2008). Several Indian startups have made notable innovations in the waste management sector.

The subsequent section describes three case studies of green ventures, which are focused on waste management issues. The products offered by these firms are unique and are based on the recycling concept of the circular economy model.

12.3.3   DESCRIPTION OF CASES

12.3.3.1   Ecokaari – Semi-Durable Goods from Multi-Layered Plastic Waste

The multi-layer plastic (MLP) covers used as wrappers in fast-moving consumer products are made from the fusion of diverse materials such as plastic films, aluminium foil, and adhesives. The fusion makes it challenging to separate and recycle the individual components effectively. Consequently, recycling multi-layer plastics becomes difficult and economically less feasible. The waste pickers ignore this type of waste, leading them to be dumped in landfills or incinerated. When MLPs are disposed of in landfills or burned, they can emit harmful gases, leading to air pollution. Moreover, if MLPs end up in water bodies, they can endanger aquatic life and the overall health of ecosystems. Ecokaari, a Pune-based green entrepreneurial venture, addresses the MLP issue with its innovative products.

Ecokaari products were made from discarded non-biodegradable and difficult-to-recycle multi-layered plastic bags, grocery bags, wrappers of cookies, chips, glittery gift wraps and old audio and video cassette tapes. The raw materials were first segregated based on their thickness, size and colour and washed using minimal water and biodegradable cleaner. After washing, the raw materials were sun-dried, manually cut, spun and rolled as yarn on a chakra (spinning wheel) into bobbins. These bobbins were used in the traditional handloom machine to weave fabric. The design team and tailors transformed the material into a range of aesthetically appealing products with a utility value. The final products included yoga mats, bags, wallets, table mats, cutlery kit pouches, cushions, and table covers. The whole manufacturing process was done manually, involving high levels of craftsmanship. The products went through stringent quality checks before hitting the market. Since the entire production process of cleaning, spinning, weaving and cutting was done manually, the firm created craft-based livelihood opportunities for women from marginalised communities.

The products were sold through the exhibitions and company website and made available in the international markets through export partners. The sales and marketing team contacted corporates, wholesalers and retailers to customise the products. Ecokaari spent considerable time creating public awareness about its processes and products. The venture conducted awareness campaigns to schools, colleges, corporations and housing societies. It regularly conducted workshops for those interested in understanding the production process.

Corporate collaboration is essential for green entrepreneurs for several reasons. Partnership with corporations provides green entrepreneurs access to resources such as funding, expertise and technology. By working with large brands, green entrepreneurs can gain more exposure and visibility, which can help grow their businesses and increase their impact. Organisations like the Tata group, Dell, Callison Flavors India, Global Connect Travels and the National Association of Software and Service Companies (NASSCOM) tied up with EcoKaari to source products for corporate gifting. EcoKaari, under its WoW Project, partnered with the Indian FMCG company ITC Limited to launch Terra by YiPPee! Products in 2023. The project aimed to provide solutions to sustainably manage post-consumer packaging waste by upcycling the waste plastic using production methods adopted by EcoKaari. In partnership with the JSW foundation, it set up an upcycling project – The Plastic Offset and Livelihoods Generation Program in Vijaynagar, Karnataka, in 2022. The mission of this collaboration was to empower seventy-five women from underprivileged backgrounds and to upcycle at least 4000 kgs of waste plastic monthly. EcoKaari received INR 50 Lakhs in funding under the Startup India Seed Fund Scheme. Ecokaari Products are shown in Figure 12.1.

12.3.3.2 Graviky Labs – Ink from Polluted Air (Recycling Carbon into Safe, High-Quality Ink)

Air pollution due to the emission of greenhouse gases (GHG) is a significant threat to human health and the environment, causing millions of premature deaths each year. GHG emissions from combustion processes, such as burning fossil fuels, contribute to releasing particulate matter (PM). PMs pose significant risks to human health, as their inhalation can lead to respiratory and cardiovascular problems.

Graviky Labs, a Bengaluru-based startup, addresses air pollution with its innovative product Air Ink. The product is made from recycled air pollutants. The Air-Ink technology uses particulate matter such as PM2.5 as raw material and microscopic particles circulated in air pollution that pose significant health risks.

FIGURE 12.1 Ecokaari products.

Fossil fuel-based vehicles and industrial exhaust systems emit these particles into the atmosphere. Graviky Labs has developed a cylindrical device called Kaalink that can be attached to the tailpipes of vehicles, and exhaust pipes. The Kaalink device captures unburned carbon or soot before it is released into the air. The captured carbon is then treated with chemicals to remove heavy metals and other harmful substances in the soot. The purified carbon is transformed into a high-quality ink pigment that can be used for various applications, including printing.

In collaboration with sustainable fashion brand Pangaia, Graviky Labs introduced the world's first apparel printed with air pollution in 2021. They have also developed applications to assist partners like Dell and MasterCard adopt more sustainable printing practices. Air-Ink has initiated a widespread public art movement across various cities worldwide, including Hong Kong, London, New York, Mumbai, Berlin, Singapore, Delhi, and more. Air pollutants have been transformed into captivating artwork, showcased by over a thousand artists such as Doodle Man, Christian Furr, Karatoes, Imagine, and many others. The innovative recycling technology employed by Air-Ink (shown in Figure 12.2) has established a unique and unprecedented business model that operates on an industrial scale. Their proprietary technology to sequester PM 2.5 and PM 10 emissions is now available for licensing and OEM integration.

12.3.3.3 Padcare Labs (Recycling of Sanitary Napkins into Cellulose)

Menstruation is a natural process in a woman's reproductive cycle where blood and tissue are discharged from the uterus through the vagina. It begins around the age of 10–12 and ends with menopause between 45 and 55 years. In earlier times, women had fewer periods due to early pregnancies, but modern women have more periods due to delayed pregnancies and early menarche. On average, a woman experiences around 400–500 periods, resulting in approximately 2000–2500 days of menstrual bleeding throughout her lifetime. Menstrual products, such as sanitary napkins, tampons, menstrual cups, and period panties, are widely used by women worldwide to manage their menstrual flow. These products can be either disposable or reusable. Plastic is the primary material in these products, with most commercial options

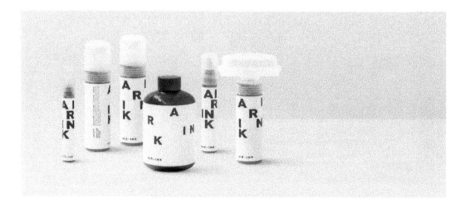

FIGURE 12.2 Graviky labs ink.

incorporating varying amounts of plastic into their design for several decades. According to the 2011 census in India, women make up 48.43% of the total population, with approximately 336 million girls and women of reproductive age who experience menstruation. Consequently, these 336 million females in India utilise menstrual products and produce menstrual waste every month. Urban areas have a higher usage rate, with nearly 70% of women using sanitary pads, compared to 48% of women in rural regions.

Although using menstrual products has significantly improved menstrual health in India, their disposal poses a growing environmental concern. The issue of menstrual waste disposal remains unaddressed, mainly due to social stigmas and taboos surrounding menstruation. Disposal practices vary based on socioeconomic status and cultural beliefs, ranging from burning the cloth to burying it in fields due to associations with witchcraft and infertility. Lack of awareness regarding the generation and disposal of menstrual waste and inadequate sanitation facilities pose a significant risk to our environment. Many menstrual product waste is improperly disposed of in water bodies, burial pits, drains, and landfills. In urban and semi-urban areas, menstrual products are wrapped in covers, newspapers, or plastic and placed in home trash or rubbish bins. Sometimes, in the lack of regular disposal facilities, garbage is wrapped and stored on rooftops of houses or beneath beds until women can figure out how to dispose of it covertly. Sanitary napkin consists of fluff pulp, polymer material, adhesives, superabsorbent material, and release paper. They were generally used and discarded like other materials. Since disposable pad contains chlorine and plastic, burning it would result in the release of toxic substances. Solid Waste Management Rules, 2016 specify that sanitary waste should be securely wrapped in the manufacturers' pouches and disposed of through incineration. However, these rules are not followed correctly, and significant sanitary waste gets mixed up with other solid waste, dumped in landfills and water bodies, or discarded openly. India produces nearly 113000 tonnes of menstrual waste annually (Toxics Link, 2021).

Padcare Labs is a circular economy startup engaged in providing menstrual hygiene management as a service to environmentally conscious B2B organisations, communities and housing societies while aligning with their sustainable development goals. The venture has developed an innovative technology that can clean, shred, and effectively decompose the harmful substances in sanitary napkins' waste and recycle them into cellulose. Padcare Labs collaborates with a network of offices and IT parks to implement environmentally safe sanitary napkin disposal methods. The used sanitary napkins are first collected in the automated safe bins in the offices. The disposable bags containing the napkins are removed from the safe bins at regular intervals and brought to their material recovery facility. The facility's recycling process uses Padcare Labs' patented technology working on 5D principles, viz., disintegration, disinfection, decolouration, deodorisation, and deactivation. The technology was developed using mechanical, chemical, and electronic features and certified by relevant government authorities. In the first stage of disintegration, the napkins, given their stickiness and elastic characteristics, are shredded using a special-purpose shredder. The shredded material is processed using a proprietary chemical solution created by Padcare Labs to disinfect, discolour and deodorise. The super absorbent polymer present in the napkin is deactivated through the processes. The final output is cellulose and plastic, which are safe to handle. The cellulose is reused to make paper products such as diaries and notepads, and the plastic is converted into granules to make padcare bins.

The client organisations of Padcare Labs are committed to ensuring their female employees' health, hygiene and wellness. In 2023, Padcare Labs served 289 clients across six cities in India recycling millions of pads. The venture's offer is modelled as Product-as-a-service to organisations with more than 500 women on a subscription model. The used pads were collected using 5000 bins placed in the washrooms of the offices. Their material recovery facility in Pune can recycle up to 50,000 pads daily. The facility can recover 99% of the material with 95% quality at $1/10^{th}$ the cost of incineration, the traditional method of disposing of the used pads. The company generates revenue from two sources. The client organisations pay subscription fees for menstrual hygiene management services under the product-as-a-service model. The client organisations are motivated to collaborate with Padcare Labs to achieve their own Zero Waste to Landfill, Environment-Societal-Governance, and gender diversity and inclusivity goals. In addition to subscription fees, the venture generates revenue from those products made from recycled plastic and cellulose materials obtained from material recovery facilities. It explores the opportunity to generate revenue under a carbon exchange model (Figure 12.3) with multinational companies. The MNCs, such as P&G and J&J, the manufacturers of sanitary napkins, are obligated to fulfil extended producer responsibility rules. Padcare Labs can help the MNCs meet their EPR using carbon exchange systems.

Green innovations such as Ecokaari, Air-Ink and Padcare Labs resulted from the conducive ecosystem prevailing in the country. India is at the cusp of making great strides in tackling waste management issues using circular economy principles. The government and mainstream businesses encourage green startups at every step in their evolution. The following section analyses the opportunities leading to the growth of green innovations in India.

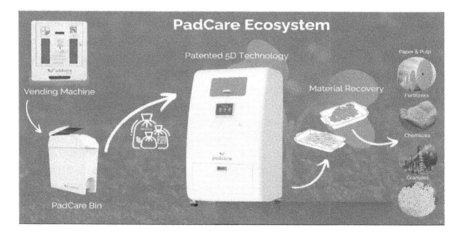

FIGURE 12.3   Padcare labs recycler.

12.4 DISCUSSION

12.4.1 Opportunities for Green Innovations

12.4.1.1 Stringent Environmental Regulations and Policies

The market for green innovations thrives on stringent environmental regulations and policies. These regulations and policies limit pollution, emissions, resource consumption, and waste generation and promote sustainable practices across various sectors. Over the past twenty years, several declarations and regulations have been aimed at protecting the environment. These measures have included limitations on chlorofluorocarbon usage, as advised by the Montreal Protocol in 1987, restrictions on CO_2 emissions, as recommended by the Kyoto Protocol in 1997, and European Community directives regarding the use of hazardous substances and electronic waste management since 2006. These regulations have increasingly impacted industries, necessitating companies to adopt more environmentally friendly practices and products (Dangelico & Pujari, 2010).

The Government of India (GoI) is committed to managing waste in the country through various initiatives, policies and regulations. The Clean India Mission is a nationwide cleanliness campaign started by the GoI in 2014 to enhance solid waste management practices and encourage cleanliness and hygiene. The campaign brought behavioural changes in people regarding cleanliness and waste disposal practices. The Solid Waste Management Rules 2016 implemented by GoI provides a comprehensive set of guidelines to manage different types of waste in environmentally responsible methods. The rules cover the responsibilities of waste generators in waste segregation, processing and disposal. Under the extended producer responsibility clause, the rules expect manufacturers, brand owners, and importers to collect and manage post-consumer waste. Since regulations related to environmental protection are made strict by governments, they create induced demand for green products and services. The extended producer responsibility requirement makes large manufacturers who hitherto operated under linear economy push to engage green ventures such as Padcare Labs to fulfil their responsibilities.

12.4.1.2 Carbon Credit and Trading Scheme

Carbon credits and trading schemes incentivise green ventures to innovate, adopt sustainable practices, and contribute to the fight against climate change. A carbon credit is a mechanism utilised to decrease carbon dioxide or greenhouse gas emissions from a project or manufacturing process undertaken by any entity. The Kyoto Protocol specifies that each carbon credit is equivalent to reducing one ton of carbon dioxide. Carbon credit and trading scheme is a worldwide strategy to combat global warming and its consequences. The process involves limiting the total emissions allowed for each organisation. If an organisation emits fewer gases than the permitted limit, the resulting shortfall is given a monetary value and can be traded under the carbon credits trading scheme. To enhance the trading of carbon credits within India, the Ministry of Power has introduced the Carbon Credit Market Scheme. The ministry issued a gazette notification on June 28, 2023, officially approving the establishment of India's inaugural domestic regulated carbon market. Carbon trading schemes encourage collaboration between green ventures and mainstream organisations. Green ventures can use these credits in regions with carbon emission regulations to meet emission reduction targets more economically. By purchasing carbon credits from green ventures, companies can offset emissions, promote sustainability efforts, and avoid penalties and reputational risks associated with non-compliance.

12.4.1.3   Access to Funding and Support

The significance of sustainability and backing green enterprises are gaining recognition among governments, investors, and organisations. Financial opportunities, grants, subsidies, and incentives are becoming more prevalent to aid the progress and expansion of environmentally conscious ventures. The Indian government offers grants and subsidies to promote green ventures and sustainable initiatives. These financial incentives support various aspects of green businesses, such as research and development, technology adoption, and renewable energy projects. For example, BIRAC – Biotechnology Industry Research Assistance Council is an initiative founded by the Ministry of Science and Technology's Department of Biotechnology in India to aid and expedite the transformation of innovative concepts and research findings into practical and beneficial products and services for society. It focuses on closing the divide between research and development and the commercialisation process by offering financial and non-financial support to startups, small and medium enterprises (SMEs), and entrepreneurs operating in the biotechnology industry. The Startup India Seed Fund Scheme is another program initiated by the Indian government to offer financial assistance to startups during their initial phases. Its purpose is to encourage entrepreneurship and innovation by providing seed funding to inventive and scalable startup concepts in diverse industries.

12.4.1.4   Demand for Green Products and Services from Environmentally Responsible Mainstream Organisations

Mainstream organisations today are committed to achieving sustainability goals. Using the Environment-Society-Governance (ESG) framework in the annual reports has gained significant importance in recent years as stakeholders increasingly consider the broader impact of businesses beyond financial performance. By integrating ESG considerations into decision-making processes, organisations aim to promote long-term sustainability, mitigate risks, and enhance their reputation. The environmental footprint is measured diligently, and green initiatives are taken to decrease greenhouse gas emissions, conserve resources, limit pollution and waste generation, and support the preservation of biodiversity. Such organisations are called "green businesses" as they adopt environmentally sustainable practices after realising the cost, innovation, marketing advantages, and ethical arguments. The growth of Padcare Labs can be attributed to the demands made by green businesses. The clients of Padcare Labs exemplify such types of green businesses.

12.5   CONCLUSION

Green innovations for circular economy embrace many sustainable practices and technologies to reduce waste, conserve resources, and promote a closed-loop system. Innovations in product design can focus on creating repairable items that can be easily disassembled, refurbished, and reused. Recycling technologies can efficiently process waste materials extending their value. Green innovations involving the development of bio-based materials that are renewable and bio-degradable can reduce the reliance on fossil fuel-derived materials and minimise environmental impact. Innovations in waste to energy present an appealing solution for environmental protection and serve as an alternative to landfill disposal of vast amounts of municipal solid waste. By converting waste into clean and dependable energy from renewable sources, they have the potential to decrease reliance on fossil fuels and curb greenhouse gas emissions resulting from waste combustion. These advanced technologies reduce waste volumes and generate significant energy output while minimising water and air pollution. The growth of green ventures in an economy is contingent upon many factors. The stringent environmental rules along with support from the government, provide the much-needed impetus in the growth of green ventures. Since mainstream organisations are increasingly held account-able for their environmental performance, they are and motivated to collaborate with green-green businesses that are adept in addressing challenges in the waste management domain. Although adopting sustainable practices can result in cost savings over the long term, certain green technologies or eco-friendly materials might entail higher initial expenses. Consequently, green businesses must thor-oughly assess the financial ramifications and balance the potential advantages against these upfront investment costs. Green-green businesses need to create awareness among the stakeholders to increase their acceptance. Green ventures invest their resources to develop and leverage a strong network with stakeholders and create awareness among potential customers to achieve a transition towards a circular economy.

REFERENCES

Borowy, I. (2013). *Defining sustainable development for our common future: A history of the World Commission on Environment and Development (Brundtland Commission).* Routledge, UK.

Charter, M. (2018). Circular economy innovation and design. In Charter, M. (Ed.) *Designing for the circular economy* (pp. 23–34) Routledge. 10.4324/9781315113067

Dangelico, R. M., & Pujari, D. (2010). Mainstreaming green product innovation: Why and how companies integrate environmental sustainability. *Journal of Business Ethics, 95,* 471–486.

Dul, J., & Hak, T. (2007). *Case study methodology in business research.* Routledge, UK

Ellen MacArthur Foundation. (2012). Towards the circular economy: Economic business rationale for an accelerated transition. https://archive.ellenmacarthurfoundation.org/assets/downloads/publications/TCE_Ellen-MacArthur-Foundation_26-Nov-2015.pdf

Falck, O., & Heblich, S. (2007). Corporate social responsibility: Doing well by doing good. *Business Horizons, 50*(3), 247–254.

Ghisellini, P., Cialani, C., & Ulgiati, S. (2016). A review on circular economy: The expected transition to a balanced interplay of environmental and economic systems. *Journal of Cleaner Production, 114*, 11–32.

Isaak, R. (2016). The making of the ecopreneur. In Making ecopreneurs. Routledge. In Schaper, M. (Ed.). *Making Ecopreneurs: Developing Sustainable Entrepreneurship* (2nd ed.). (pp. 63–78) Routledge, UK. 10.4324/9781315593302

Kemp, R., & Pearson, P. (2008). *Policy brief about measuring eco-innovation and Magazine/ Newsletter articles.* Um Merit, Maastricht. https://cordis.europa.eu/docs/results/44/ 44513/124548931-6_en.pdf

Kenway, J., & Fahey, J. (2014). Is greed still good? Was it ever? Exploring the emoscapes of the global financial crisis. In *Education, Capitalism and the Global Crisis* (1st ed.). (pp. 17–27). Routledge, UK. 10.4324/9781315872308

Meena, M. D., Dotaniya, M. L., Meena, B. L., Rai, P. K., Antil, R. S., Meena, H. S., ... & Meena, R. B. (2023). Municipal solid waste: Opportunities, challenges and management policies in India: A review. *Waste Management Bulletin, 1*(1), 4–18.

Planning Commission Report. (2014). Retrieved from https://www.niti.gov.in/sites/default/ files/2018-12/Niti_annual_report-2014-15.pdf

Toxics Link (2021). Menstrual Products and their disposal, New Delhi. https://toxicslink.org/ wp-content/uploads/2022/08/Menstrual%20Waste%20Report.pdf

Walter, R. (2016). Stahel circular economy. *Nature,* 6–9. 10.1038/531435a

Webster, K. (2021). A circular economy is about the economy. *Circular Economy and Sustainability, 1*(1), 115–126. https://link.springer.com/article/10.1007/s43615-021-00034-z

Yosef, J (2008). A new conceptual framework for sustainable development. *In Environment, Development and Sustainability: A Multidisciplinary Approach to the Theory and Practice of Sustainable Development, Springer,* 10(2), 179–192, April. 10.1007/s1 0668-006-9058-z

Yadav, S., Samadhiya, A., Kumar, A., Majumdar, A., Garza-Reyes, J. A., & Luthra, S. (2023). Achieving the sustainable development goals through net zero emissions: Innovation-driven strategies for transitioning from incremental to radical lean, green and digital technologies. *Resources, Conservation and Recycling, 197,* 107094. 10.101 6/j.resconrec.2023.107094

Index

Ability-Motivation-Opportunity (A-M-O), 6
Affective Component, 73
Altruism, 10, 11
Architecture Design, 125
Artificial Intelligence (AI), 33
Asian Market, 73
Attitude, 73, 74, 75, 76, 77, 78

Biodegradable Crafts, 44
Biodegradable Polymers, 34
Biodegradable Substances, 29
Brand Attachment, 29
Brand Image, 58, 66

Canonical Correlation Analysis (CCA), 5, 6, 8,
        9, 10
Canonical Variate, 9
Carbon Credit, 128, 138, 173, 175
Carbon Footprints, 112, 113, 156
Carbon Footprints, 60, 62
Chena Cultivation, 47
Circular Economy, 25, 27, 28, 31, 32, 34, 166,
        167, 168, 169, 170, 171, 173, 174,
        175, 177
Circular Economy, 87, 88, 93, 99, 102, 103
Circular Economy and Innovation, 102
Circular Economy model, 154
Circular Economy Package, 25, 27
Clean Transportation, 132, 134
Climate Change (IPCC), 153
Climate Change Finance Unit (CCFU), 136
CO2 Emission, 128
Cognitive Component, 73
Common Threads Garment Recycling
        program, 32
Competitive Advantage, 57, 58, 59, 60, 92
Conservation Technologies, 122
Consumer Behaviour, 73, 75
Contemporary Antecedents, 1
Cooperative Research Projects, 124
Corporate Greening Model, 62
Corporate Social Responsibility (CSR), 133, 136
Cosmetic Industry, 72
Cultural Heritage, 44
Customs, 40, 41, 43, 47, 48, 53
Cyclable Packaging, 2

Dambana Indigenous Community, 46
Department Of Economic And Policy Research
        (DEPR), 127

Developing Countries, 72, 77, 78
Digital Twins, 155, 156, 157, 158, 159, 160, 161,
        162, 163
Digital twin implementation framework, 154, 156

Eco Friendly, 47
Eco-Friendly Behaviour, 3
Eco-Friendly Entrepreneurship, 47
Ecokaari, 167, 170, 171, 174
E-Commerce, 22, 23, 24, 25, 26, 27, 28, 29, 30,
        31, 32, 34, 35
E-commerce Green Packaging, 29
economic Ramifications, 41
Eco-Tourism, 44
Electric Vehicles, 125
Electronic Word of Mouth (eWOM), 73
Electronic Word of Mouth, 73, 74
Embrace Energy Efficiency Practices, 104
Emerging Markets, 78
Energy Efficiency, 122, 124, 125
Energy Efficient Projects, 130, 132
Energy And Resource Efficiency, 100
Entrepreneurial Challenges, 44
Entrepreneurial Practices, 43, 48, 53, 54
Environment Friendly technologies, 132
Environment-Society-Governance, 137
Environmental Attitude (EA), 4, 10
Environmental Awareness, 83
Environmental Entrepreneurship, 47
Environmental Footprint, 28, 29, 30, 35
Environmental Impact Reduction, 98
Environmental Issues, 120
Environmental Justice, 123
Environmental Performance, 57, 58, 59, 60, 61,
        63, 64
Environmental Problems, 72
Environmental Risk Management System, 95, 97
Environmental, Social And Governance (ESG),
        136, 137, 142
Environmental Sustainability, 57, 58, 133
European Union Packaging and Packaging, 24, 27
Extended Producer Responsibility, 174, 175

Family Controlled Business (FCB), 57, 58, 59, 60,
        62, 65
Family-Run Hospitality Business, 62, 64, 65, 67
Fossil Fuels, 128, 129, 142
Fourth Industrial Revolution, 155
Fractal structures, 151
Fractals, 146, 149, 151

Graviky Labs, 167, 171, 172
Green Awareness (GA), 3, 10
Green Banking, 133
Green Behaviour (GB), 2, 3, 10
The Green Bond Principles, 134, 138, 139,
    140, 141
Green Bonds, 128, 130, 131
Green Bonds Principles (GBP), 138
Green Building, 124
Green Business, 177
Green Business Value Chain, 65
Green Competencies Training, 66
Green Concerns, 73, 75, 76, 77, 78
Green Cosmetics, 72, 73, 74, 75, 76, 77, 78
Green Culture (GC), 3, 10, 11
Green Culture, 61
Green Economy, 132
Green Employee Engagement, 58, 62, 64
Green Employee Retention, 58, 62, 63, 64, 66
Green Entrepreneurship, 41, 44, 46, 47, 48, 49, 52
Green Factors, 5
Green Financing, 133, 134
Green Habit (GH), 3, 10, 11
Green House Gas (GHG), 127
Green HR, 57, 59, 61, 63, 65
Green Innovation, 44, 54, 58, 62, 64, 65, 67,
    167, 170
Green Innovation Practices, 53
Green Manufacturing, 84, 89, 95, 97, 154, 155,
    156, 157, 159, 160, 161, 162
Green Marketing, 5, 12
Green Mortgages, 133
Green Packaging, 23, 24, 25, 26, 27, 28, 29, 30,
    31, 32, 33, 84, 95, 88, 97, 99
Green Performance Management System
    (GPMS), 58, 62, 63, 66
Green Production, 154, 156, 160, 161
Green Products, 41, 47, 52, 53, 73, 74, 75, 77
Green Purchase Behaviour, 10
Green Purchase Intention (GPI), 73, 74, 75, 76,
    77, 78
Green Purchase Intention, 10
Green Purchasing, 96, 97, 103
Green Purchasing Decisions, 5
Green Recruitment And Selection, 58, 62, 63, 66
Green Retailing, 96, 97
Green Risk Management, 58, 62, 65
Green Supplier, 95, 97
Green Supply Chain, 82, 83, 84, 85, 86, 89, 91, 92,
    93, 102, 103, 104, 105
Green Supply Chain Management, 57, 89, 91, 92,
    93, 94, 102, 103, 104, 105
Green Training And Development, 58, 61, 62, 63
Green Washing, 56

Green Work Life Balance (GWLB), 57, 58, 62,
    63, 66
Greenhouse Gas (GHG), 27
Gross Domestic Product (GDP), 127

Hospitality Industry, 56, 57, 59, 60, 62
Human Wellbeing, 127

India INX, 143
Indian Green Consumer Behaviour, 72
Indian Green Consumption, 11
Indian Millennials, 4
Indian Renewable Energy Development Agency
    Ltd, (IREDA), 137
Indigenous Community, 40, 41, 42, 43, 44, 45, 46,
    47, 48, 49, 50, 51, 52, 54
Indigenous Entrepreneurship, 46, 52
Indigenous Knowledge, 44, 45, 52
Indigenous Medicines, 52
Indigenous People, 43, 44, 45, 46, 47, 52
Indigenous Resources, 48
Indigenous Tourism, 51, 52
Indigenous Vedda Community, 40
Industrialization, 127, 130
Innovations in Waste Management, 122
Innovative Green Business Models, 44
Intelligent Packaging Technology, 33
Internal Structure Optimization, 134
International Labour Organization (ILO), 128
Internet of Things (IoT), 89, 125
Interpersonal Influence, 3, 10
Irresponsible Consumption, 77

Kaalink, 172
key Performance Indicators (KPI), 102
Knowledge Sharing, 124

Life Cycle Thinking, 94, 121
Lower Middle-Income Countries, 138

Marketing Communications, 73, 78
Mediating Role of Attitude, 73, 76, 77
Mediating Role of Green Concerns, 73, 76, 77
Methodology - Prisma, 83
Multi-Layered Plastic bags, 170
Multiple Linear Regression, 5

Nanotechnology, 27
National Solar Mission, 133
National Thermal Power Corporation
    (NTPC), 137
Natural Resource, 43, 44
Non Banking Financial Companies (NBFC), 135
Non Governmental Organizations (NGO), 88

Noor Iteration, 149, 151

Optimize Logistics and Transportation, 103

Padcare Labs, 167, 173, 174, 175, 176
Paris Agreement, 124
Pearson Correlation, 5, 6, 10
Perceived Environmental Knowledge, 4, 10
Performance Measurement and Reporting, 101
Physical and Digital Worlds, 154
Polyethylene Terephthalate (PET), 27
Private Limited (PVT), 51
Process Digital Twins, 155

Rathugala Indigenous Community, 49, 52
Recycling, 5, 11
Recycling Technologies, 122
Regulatory Framework, 124, 125
Renewable Energy, 122, 123, 124, 125
Renewable Energy Projects, 131, 133, 135, 137
Renewable Energy Sources, 121, 122, 123, 124, 125
Reserve Bank Of India (RBI), 137
Resource-Based Theory, 59

Security and Exchange Board of India (SEBI),
        136, 137, 138
Self-Similarity, 149
Significant Factors, 2
Small and Medium Enterprise (SME), 62
Smart Green Technology, 120, 121, 122, 123,
        124, 125
Social Cognitive Theory (SCT), 59, 61
Social Entrepreneurship, 47
Social Equity, 123
Social Media Marketing, 4, 10, 11
Social Repercussions, 123
Solar Energy, 151
Solar Energy Efficiency, 122, 124, 125
Solar Panel, 146, 147, 149, 150, 151
Solid Waste Management, 172, 175
Source Creditability, 74
Sovereign Green Bonds, 137, 142
Sri Lanka, 40, 41, 42, 43, 44, 45, 46, 47, 48, 50, 52
Stakeholder Engagement and Transparency, 101
Stakeholder Theory, 54, 60
Stakeholders, 82, 92, 96, 97, 98, 99, 101, 102,
        103, 104
Startup India Seed Fund Scheme, 176

Stimulus, 77
Stimulus Organism Response, 73
Subjective Norms, 11
Supervisory Control and Data Acquisition
        (SCADA), 159
Sustainability, 169, 170, 172, 174, 176
Sustainability, 43, 44, 45, 47, 48, 50, 52
Sustainable Business Model, 41, 54
Sustainable Development, 24, 37, 167, 170, 173
Sustainable Development Goals, 156
Sustainable Economic Growth, 128
Sustainable Harvesting, 41
Sustainable Packaging, 28, 33, 35
Sustainable Resource Management
        Techniques, 41
Sustainable Sourcing and Material
        Management, 99
Sustainable Transportation, 121
System Approach, 122
System Digital Twins, 155

Technological Transfer, 124
Theory of Planned Behaviour (TPB), 73
Theory of Reasoned Action (TRA), 73
Theory of Reasoned Action, 73, 75
Tourism and Hospitality Sector, 66
Traditional Practices, 51, 52, 53
Triple Bottom Line, 47

Unique Selling Proposition (USP), 60
United State Dollar (USD), 141
Upper Middle-Income Countries, 130

Vakarei, 47

Wakare Indigenous Community, 41, 43
Waste Management, 121, 122, 123, 125, 170, 173,
        174, 175
Waste Reduction And Recycling, 93, 94
Wasteful Packaging, 23
Water Conservation, 122
Word of Mouth (WOM), 73, 74
Workplace Hazards, 65, 66
World Economic Forum, 153

Young Generation, 53

Zero Waste, 174

Milton Keynes UK
Ingram Content Group UK Ltd.
UKHW031133141024
449569UK00006B/231